FUTURE
MINDS

Also by Richard Yonck

*Heart of the Machine: Our Future in a World
of Artificial Emotional Intelligence*

FUTURE
MINDS

THE RISE OF INTELLIGENCE
FROM THE BIG BANG
TO THE END OF THE UNIVERSE

RICHARD YONCK

Arcade Publishing • New York

First Edition

Arcade Publishing books may be purchased in bulk at special discounts for sales promotion, corporate gifts, fund-raising, or educational purposes. Special editions can also be created to specifications. For details, contact the Special Sales Department, Arcade Publishing, 307 West 36th Street, 11th Floor, New York, NY 10018 or arcade@skyhorsepublishing.com.

Arcade Publishing® is a registered trademark of Skyhorse Publishing, Inc.®, a Delaware corporation.

Visit our website at www.arcadepub.com.
Visit the author's site at www.intelligent-future.com.

10 9 8 7 6 5 4 3 2 1

Names: Yonck, Richard, author.
Title: Future minds: the rise of intelligence, from the big bang to the
    end of the universe / Richard Yonck.
Description: First edition. | New Yor: Arcade Publishing, [2020] |
    Includes bibliographical references.
Identifiers: LCCN 2019050009 | ISBN 9781948924382 (hardback) | ISBN
    9781948924405 (ebook)
Subjects: LCSH: Artificial intelligence—Philosophy. | Humanity.
Classification: LCC Q334.7 .Y66 2020 | DDC 006.301—dc23
LC record available at https://lccn.loc.gov/2019050009

Cover design by Erin Seaward-Hiatt
Cover illustration: © Sololos/Getty Images (cosmos); © PeopleImages/Getty Images
(woman's face)

Printed in the United States of America

For my favorite intelligence in the universe—Alex

"'Now' is never just a moment. The Long Now is the recognition that the precise moment you're in grows out of the past and is a seed for the future."

—Brian Eno

CONTENTS

CONTENTS

PREFACE

We are all time travelers.

So many things make human intelligence unique. Our use of language and tools. Our ability to transform knowledge and concepts into entire sciences or schools of thought. Our refined motor skills that allow us to grasp everything from a soap bubble to a sledgehammer. But of all these wondrous abilities, none is so uncanny as our ability to travel through time. It is perhaps our greatest gift, and in so many ways it makes us what we are: *Homo tempus*, the time-traveling species.

You may think I'm playing with words, because in the traditional sense, science assures us that time travel is impossible. In another, more mundane sense, everything moves along the arrow of time. Whether we're talking about a rock, a rocking horse, or a rock star, all the universe passes through the seconds, minutes, and hours at the same metronomic pace. All of it consistently, in one unified direction, as we march to time's unceasing beat.

But that is not what I mean at all, because we human beings are time travelers in the truest sense of the word, flitting from past to present to future and back again as effortlessly as a butterfly flutters from one flower to the next. This amazing capacity is a gift of our minds. With it, we are able to move from moment to moment with such ease and from such an early age that we rarely see this ability for what it truly is.

At times, it may seem as though our time-traveling skill is in charge of us rather than the other way around. Perhaps one moment we're sitting in a monotonous business meeting, when suddenly we are transported to an event from our childhood decades before. Or maybe we'll be walking down the street as we relive a conversation from earlier that morning. The conditions aren't important; the result is. Our minds let us wander effortlessly through this timescape, transcending the physical laws of the universe, allowing us to visit any place, any moment our memory and imagination want to deliver to us.

It is an astounding ability, this time-travel power of ours. It allows us to anticipate what is to come and to continue to grow and learn from the moments that have long since passed. Perhaps most importantly, it makes it possible for us to endeavor to build a future—ideally our preferred future, where we hope to live one day.

Which brings us to the matter of this book about the future of intelligence, a book by and for time travelers. For this author, it is an opportunity to take a Big History view (which I'll explain momentarily), to jump about millions, even billions of years in the course of telling this story, which is thrilling beyond words. Having time-traveled this way all of my life, here is a chance to share the journey, to revel with others in the majestic beauty of our emergent and increasingly intelligent universe.

But if that were the only reason for this approach, our story would be little more than a travelogue, which I earnestly hope it is not. Instead, this book affords an opportunity to take the broadest of perspectives on intelligence, exploring the concept from a vantage that spans the bounds of the cosmos from the earliest instant to the end of all time.

In order to develop this Big History time-traveling view, *Future Minds* is organized into three sections. "Deep Past" covers the beginning of the Big Bang through the twentieth century and looks at how the laws of the universe have enabled the development and evolution of complexity, life, and intelligence and what this may mean for our future. "Twenty-First Century" explores the existing and anticipated developments in artificial intelligence, augmented human intelligence, and many other related areas. Finally, "Deep Future" extrapolates these developments and trends to speculate on the remaining 100 trillion years or more until the end of the universe. (I didn't want any epoch to feel left out!)

There will be those who take issue with some, perhaps many of the views that are presented here. Nevertheless, I believe we are at a stage in our evolution that necessitates we reinterpret and redefine that suitcase word[1] *intelligence*. Long before René Descartes wrote his famous "*cogito ergo sum*," people pondered the nature of thought and experience, trying to understand just what these things we call *intelligence* and *consciousness* are. While this book can hardly promise a resolution to millennia of study, introspection, and conjecture, it will hopefully be a stimulus for ongoing discussion.

Finally, I want to underscore that this is not just any journey but a most human journey. Regardless of how far back we determine that intelligence originated, those earliest precursors eventually led to *us*. Similarly, however this thing we call *intelligence* develops, at least in our little pocket of the universe, we will have had a role in helping it get there. And who knows? Perhaps we are destined to be part of whatever that future of intelligence becomes. But if so, we should probably ask ourselves: What does the future hold that could be greater than our mind's extraordinary ability to journey effortlessly through all of time and space? Perhaps we might find the answer in these words attributed to the incomparable Albert Einstein: "Logic will get you from A to B. Imagination will take you everywhere."[2]

DEEP PAST

DEEP PAST

CHAPTER I

INTELLIGENCE IN THE UNIVERSE: OR, WHERE IS EVERYBODY?

"The number of technological civilizations should literally number in the millions in our galaxy alone."
—Carl Sagan, cosmologist, astrophysicist, author

"But where is everybody?"
—Enrico Fermi, Nobel Prize–winning physicist

When I was a young boy, I would frequently journey deep into space, negotiating asteroid fields, solar prominences, and luminous nebula that would one day become the birthplace of countless new stars. Many years before the first astronauts set foot on the moon, I'd lift off from my bedroom at the north end of Seattle, late at night, after the rest of my family had gone to sleep. (Long before Microsoft and Amazon, this was the Seattle of lumber mills, commercial fishing vessels, and one lone industrial giant, Boeing.)

Launching from my bedroom, my ship would soar into the night sky, the stars filling my field of view as I rapidly left our world behind me. Whipping around our moon a few times for extra acceleration, I'd slingshot toward Jupiter or sometimes Saturn, seeking a similar though far greater gravity assist from one of those massive gas giants. Soon I was rocketing toward one of our nearest stellar neighbors, nearby Rigil Kentaurus (also known as Alpha Centauri), Tau Ceti, or perhaps the lyrically named Epsilon Eridani. The distances closed quickly as some exotic, now-forgotten mechanism allowed my ship's dynadrive to accelerate to velocities far greater than the speed of light.

The cosmos was vast, beautiful, dazzling, even more picturesque than the many books I regularly checked out from the library so I could study and explore our extraterrestrial backyard from the comfort of my home world. I supplemented the books with the increasing number of grainy black-and-white telecasts NASA shared from their control rooms at Houston and Cape Canaveral, which had only recently been renamed Cape Kennedy for the most heartbreaking of reasons. Long before home VCRs, I'd make screen captures with my slender 126 Instamatic camera, its diminutive form the embodiment of miniaturization we'd come to associate with modern technological progress.

While my extraterrestrial travels didn't take place every night, they were very frequent as I surveyed the ever stranger and more mysterious phenomena and formations I'd read about by day. Despite the vast distances involved (which I thought I grasped, though I'm now sure I didn't), my exotic ship easily delivered me all the way to the edge of the universe before I could drift to sleep. (The CMB— the cosmic microwave background—was still years from entering our textbooks and library shelves, and the Big Bang wouldn't be accepted over steady state theory for several more years, so fortunately my dynadrive didn't have to contend with these rather significant revelations.)

One thing that was very evident from my "travels" was the apparent absence of other forms of intelligent life in the universe. Most serious astronomy and cosmology books at the time seemed to be all but sterilized of the concept, as though the photos and information had undergone decontamination procedures as they reentered Earth's atmosphere. The publishers were making it very evident they were dealing with hard science and that any speculation about life existing elsewhere should be left to comic books, pulp science fiction shelves, and *Star Trek*.

Star Trek,[1] of course, was the other major space influence on me from that era of the mid-1960s. It acknowledged a universe rich with life of every possible variety . . . that is, until you peeked under the deflector shields and realized that nearly every alien was a bipedal humanoid who more often than not spoke perfect English with a Midwest or occasionally even a London accent. This despite any sign of a universal translator being near at hand. It was as if they were surreptitiously telling us that we were the one and only intelligence we could count on finding in this cold, vast universe as we began to take our first baby steps beyond Planet Earth. So, while my nightly sojourns might occasionally find me imagining a conversation with a tall, graceful Tau Cetian, I was far more likely to cross a dark, lifeless void. No living being came in range of my sensors.

As the space race picked up speed, it also became rapidly evident that there were no extraterrestrials hiding out from us on the dark side of the moon—neither as invaders lying in wait nor as members of some great galactic welcome wagon waiting to leap out with a universally translated shout of "Surprise!" as they received us into their vast and decreasingly exclusive cosmic club.

More time passed. From radio silence to moon rocks to Mars soil, the absence of any life-confirming discoveries only added to the sense that perhaps Enrico Fermi had been correct. The Italian American physicist originally achieved fame for developing the first nuclear reactor, for which he later won the Nobel Prize in Physics in 1938. More than a decade later, as the story goes, Fermi was working at Los Alamos having lunch with some of his esteemed colleagues (Edward Teller, Emil Konopinski, and Herbert York) when the conversation turned to alien life. The news had recently contained several stories and cartoons about UFOs, leading to speculation about the numerical likelihood of space aliens. Following a long pause, Fermi reputedly responded, "But where is everybody?"[2]

Though his question was received humorously, it succinctly addressed a contradiction that has henceforth been known as the Fermi paradox. If, as Fermi and others calculated, the universe should be teeming with non-Earth life, why had we still not been contacted or otherwise detected evidence of its existence?

Such calculations would later be famously formalized by astronomer and astrophysicist Frank Drake in 1961, in his eponymous Drake equation[3]:

$$N = R_\star \cdot f_p \cdot n_e \cdot f_l \cdot f_i \cdot f_c \cdot L$$

where:

N = the number of detectable civilizations in our galaxy

and:

$R_\star$ = the average rate of star formation per year

f_p = the fraction of those stars that have planets

n_e = the average number of planets that can potentially support life per star that has planets

f_l = the fraction of these that go on to develop life at some point

f_i = the fraction of these that go on to develop intelligent life

f_c = the fraction of civilizations that develop technology that releases detectable signs of their existence into space

L = the length of time for which such civilizations release those detectable signals

Based on these and other researchers' calculations, many people believed there should be millions of civilizations in our galaxy alone, more than justifying Fermi's response. So indeed, where was everyone? Had these hypothetical civilizations unfailingly destroyed themselves upon reaching a certain level of technological sophistication? Then again, maybe they didn't feel we were ready to meet them yet, that we weren't yet mature enough as a civilization. Or perhaps they were intentionally hiding in order to elude detection by something we were still too dumb to know we needed to avoid. Or was it possible that some as yet unknown law of physics prevented communication or travel over such vast distances?

No matter the cause, it was becoming very evident that if intelligent life did exist beyond our planet, it must be far rarer or more taciturn than many of us had hoped. Yes, the scale of even our immediate cosmic neighborhood is stiflingly vast; certainly far more so than I had ever conceived as a young boy. But shouldn't we be able to catch at least some glimpse, some hint that we aren't an absolute anomaly? That we aren't alone? By accepting our existence as so exceptional, aren't we directly refuting the mediocrity principle[4] that all but guarantees life beyond this one planet from a statistical standpoint? How ironic would it be to only now discover that after all of this time, we actually do reside at the center of the universe, singularly aware and singularly alone?

Ironic though that might be, it also remains highly unlikely. Especially given that at this stage, we've barely so much as peered beyond our own cosmological back porch. In light of the tremendous vastness of the cosmos, the discovery of extraterrestrial life may well remain a considerable challenge for a very, very long time to come.

Given all of the truly extraordinary circumstances that would be required for humankind to stand alone and make this claim to our "universal exceptionalism," perhaps what we need to do in the face of so much contradictory evidence is apply Occam's razor. We need to ask ourselves what might be a more reasonable explanation. More importantly, perhaps we need to reconsider the question entirely.

We are rapidly approaching the point when we will see the development of many types of new intelligence here on Earth. Some will be biological. Others may be silicon-based. Some will be a blend of the two, and others will be of origins we have yet to imagine. While this will no doubt lead to many changes, one major outcome will almost certainly be a growth in our understanding of our own place and purpose in the cosmos. Though many of us have believed life and intelligence would be prevalent throughout the universe, we have yet to prove it so. Indeed, if this view is wrong, the revelation will no doubt lead to significant introspection about what our place should be in the great scheme of things.

However, it may also be that we haven't been looking in the right place or in the right way. It may be that what we are seeking is hidden right before our eyes and we don't know that we're looking straight at it. Perhaps we need to consider the matter afresh in order to find what we seek in the vast, deeply woven fabric of the universe.

In order to realign our view of intelligence here on Earth and throughout the cosmos as a whole, perhaps we need to consider the possibility that we've been thinking about intelligence in entirely the wrong way. We've generally assumed alien life will be very different from ourselves, but so many of our assumptions to date appear to be based on a technological world built by beings not all so very different from ourselves, with similar evolutionary origins, technical histories, emotional motivations, and sensoriums (sense organs and related cognitive processing). But this is such a narrow definition; it doesn't even do all that good a job of describing variations among diverse groups of human beings. Using one limited example, were we to orbit an alien world, how likely would we be to discover a species of highly empathic, socially advanced, non–tool wielding transparent jellyfish-like creatures? While these inhabitants might be highly intel-ligent by a number of different measures, several factors would limit their being discovered even by an orbiting space probe, much less a remote sensor halfway across the galaxy.

Another example might be a globally connected nodal root sys-tem that gives rise to a cognitively active advanced intelligence, only this network operates on significantly slower timescales than our own brains. While it would be easy to call this network a plant or fungus given our own evolutionary history, that would be a misrepresentation of its extremely different genetic background. Such an intelligence might maintain some types of globe-spanning memories stretching back thousands, perhaps even millions of years, but be virtually without external motivation, localized perspective, or ego. (Idea inspired by author Ursula K. Le Guin.)

On the other hand, if we look to machines as our next exposure to a new intelligence, we may find them not so anthropomorphic as we expected. Yes, they may communicate with voices that seem human, and they may eventually wear faces indistinguishable from our own. Nevertheless, in time their inner worlds, their motivations, their priorities may become vastly different from those of human beings.

All intelligences needn't mimic human intelligence to be considered of a high or perhaps even "higher" order. Depending on a range of factors, an intelligence could be lacking in many qualities we consider essential to human intellects and still be vastly superior to ourselves in other areas. Under such circumstances, which is the greater intellect? According to whose perspective and criteria? More importantly, is this even a valid comparison to make as we move into an era of rapidly developing new intelligences?

Ultimately, the purpose of this speculation is to set the stage for exploring the nature and future of intelligence. At times, it may be that what we find will be very recognizable to us as intelligence. On the other hand, it may be that time and again we run up against something we wouldn't ordinarily categorize as intelligence at all, despite its being vastly superior to humans, taken from some different perspective. Which is exactly why we should be challenging our assumptions about intelligence in the first place.

There may well be many means for addressing and categorizing such differences; methods that transcend culture, species, or even morphology. In this initial section, we will explore the physical laws, conditions, and stimuli from which intelligence arises, in order to identify those factors that are universal to its development, regardless of what form it may eventually take. Ideally, this will help us to better understand and recognize intelligence in all its potential forms, be it terrestrial or alien, biological, technological, or of origins we have yet to encounter or even imagine.

As we begin what's likely to be a long and winding journey, let's start by asking ourselves one very essential question.

What Is Intelligence?

Are you more intelligent than an ant? I would wager that most people answering seriously would immediately say "Yes." Now consider the question "Based on what criteria?" and things get a tad more interesting. Many of us will point to our use of tools and accomplishments as a species. Others may hold up our mastery of syntactic language and our ability to compose sonnets and arias. The list would go on for a long time.

Now consider yourself an ant. Forgive my anthropomorphizing, but you are asked the same questions about those great hairless apes you occasionally see lumbering about, mindlessly destroying your hard-built structures. These are your sonnets and arias, these underground palaces with their labyrinthine passages and chambers. Your mandibles and your many-sectioned legs are your tools, ready whenever you need them. Moreover, it is evident by their clumsy diggings that these apes have no understanding of the pheromonal and low-resonance language you use to share information with the rest of your kind. Besides, ants have been here for over a hundred million years. By comparison, these dumb brutes only arrived yesterday.

Obviously, this example is asking a lot of these industrious insects, but what about if we turn the scenario around? You're a human being again—an astronaut—and you're asked to assess a set of stones located on a small, lifeless planetoid in the distant reaches of a nearby star system. To the untrained eye, these stones look much like common basalt and display nothing immediately unusual except for their isolated location. They seem to be nothing more than dumb rocks.

However, after following the appropriate protocols for removing samples to your ship, it quickly becomes evident you've made a very bad mistake. Orbiting at the far reaches of its solar system, the planetoid is virtually without an atmosphere and therefore its ambient temperature is approximately 40° above absolute zero. But on removal to the storage bins where samples are stored on the outside of the ship, at a sweltering minus 100°C (173° above absolute zero), the stones begin to glow and pulse rapidly. Before your crew can react, the stones go critical, releasing so much energy in a fraction of a second that it wipes out the entire solar system, including your ship and crew along with it.

The stones weren't dumb rocks, but were in fact pure computronium, a hypothetical form of programmable smart matter designed to maximize the computational capability of every one of its atoms. The gravitational flux of a mini black hole at the center of each stone acted as a power source that had already lasted a billion years. Waste heat was mostly limited by the process of reversible computing, which negated the supernova-scale temperatures generated by the vast computing power of the stones. The remaining residual heat was radiated away by the carefully balanced environment of their seemingly lifeless world.

And what were the stones computing? Only the lives and realities of more than a trillion virtual residents who had established the stones there, a billion years before. To suggest these virtual lives weren't real would only be an indication of how primitive we still are in our thinking about intelligence. This had been one of the earliest civilizations in the universe. When it established its virtual society, there were no signs of other life or intelligence within their light cone (anywhere within their observable universe), and so at the time the risks of locating the civilization-bearing computronium were deemed minimal—threatened only by the rare possibility of colliding with an asteroid. Perhaps, had the human explorers taken note of other anomalies, such as the historic deflection paths of all asteroids and meteors in that part of the system, they would have realized intelligence had long been at work there with a purpose. Instead, they wiped out themselves and more than a trillion citizens of a billion-year-old civilization.

———————

What is intelligence? It seems a basic enough question, doesn't it? Intelligence is our ability to think—broadly, deeply, inquisitively. It is what allows us to make decisions, the tool that helps us to solve problems.

Yet there are so many aspects of intelligence that have nothing to do with any of these. Our ability to appreciate a crimson rose or to be repulsed by a fetid odor. To be moved by a sunset or a piece of music. To find joy in the burble of a baby. Intelligence allows us to distinguish right from wrong, to create works of art, to contemplate the wonders of the universe.

Intelligence allows us to ponder the nature of intelligence. Yet, as broad as all of these descriptions are, they barely begin to scratch the surface. In many ways, they may be leaving far too many forms and processes that should qualify as intelligence unrecognized.

Increasingly, these days, we find various groups who are willing to extend the label of intelligence to include minds beyond our own. For instance, many biologists and cognitive scientists believe certain animals have the capacity to experience phenomena in the world much as we do and therefore are deemed to have some degree of intelligence, if not also consciousness. On the other end of the continuum, a large number of technology researchers and experts think computers that implement certain types of artificial intelligence may come to share at least some of these traits with us over time.

Consider Thomas Nagel's 1974 philosophical treatise, "What Is It Like to Be a Bat?"[5] While the paper's primary focus is on the irreducibility of consciousness, it also makes a strong case for the impossibility of accurately conveying and sharing subjective phenomena in general. By this reasoning, the inner world of any entity, but especially of another species, is all but unknowable to us. Because so many aspects of our intelligence are deeply entwined with our subjective world of consciousness, this could create a barrier not just to understanding but to recognition as well.

The reality is that even within our limited microcosm of Earth, there may be a gamut of intelligences we will never fully appreciate or perhaps even recognize. The reason for this may lie in how we think about it. In looking at attempts to formally define intelligence in the general literature, we find most focus on reasoning, abstract thought, and other "higher" cognitive functions. For instance, considering a typical definition of intelligence from an encyclopedia, we find:

> The general mental ability involved in calculating, reasoning, perceiving relationships and analogies, learning quickly, storing and retrieving information, using language fluently, classifying, generalizing, and adjusting to new situations (*Columbia Encyclopedia*, 6th edition).

This shortchanges a number of aspects of our own intelligence that aren't directly tied to those processes involved in abstract and analytical thought. Such definitions would also exclude nearly all nonhuman animals, a situation that many animal cognition researchers would certainly take issue with. As Nagel points out, it would be very presumptuous of us to try to explicitly define what bat intelligence is like. All we can do is broaden our definition in a way that allows at least some, if not all, animals to be included. While animals cannot perform calculus, much less invent it, there are any number of ways many species can use their intellects to solve problems, to use tools, to experience the world.

Just as with bats navigating by echolocation, there are many ways in which different animals are intellectually superior to humans in their own right. Therefore, the following declaration may be considerably closer to what we are looking for. Formally known as the "Cambridge Declaration on Consciousness," it was made by a prominent international group of cognitive neuroscientists, pharmacologists, neurophysiologists, neuroanatomists, and computational neuroscientists gathered at the University of Cambridge in 2012.

> The absence of a neocortex does not appear to preclude an organism from experiencing affective states. Convergent evidence indicates that non-human animals have the neuroanatomical, neurochemical, and neurophysiological substrates of conscious

states along with the capacity to exhibit intentional behaviors. Consequently, the weight of evidence indicates that humans are not unique in possessing the neurological substrates that generate consciousness. Non-human animals, including all mammals and birds, and many other creatures, including octopuses, also possess these neurological substrates.

Though we are discussing intelligence as opposed to consciousness, it seems reasonable to say that should something be capable of experiencing consciousness, it must therefore have some significant degree of intelligence as well. If we consider consciousness as an internal response to events and experiences, both internal and external, then even by the more restrictive definitions, this should still count as a form of intelligence.

The reverse is not necessarily true, however. There is nothing apparently inherent in consciousness that should make it a requisite of intelligence. Indeed, we may find that there are many members, or even branches, of the animal kingdom that don't meet any of our definitions of consciousness but that nonetheless exhibit certain degrees of intelligence. Therefore, it would seem that our definition needs to broaden further, allowing for nonhuman and even nonconscious intelligence.

Which brings us to nonbiological intelligences. While we may yet determine that nonbiological systems cannot achieve consciousness or experience affective sensations in any true sense, this hardly precludes them from exhibiting intelligent behavior. A few definitions from some notable artificial intelligence researchers attempt to broaden our views on intelligence.

"Intelligence is the ability to use optimally limited resources—including time—to achieve goals." (Ray Kurzweil, inventor, director of engineering, Google)

"Any system . . . that generates adaptive behaviour to meet goals in a range of environments can be said to be intelligent." (David Fogel, AI engineer, Lockheed Martin)

> "Intelligence may be defined as the ability to achieve complex goals in complex environments using limited resources." (Ben Goertzel, AI researcher, chief scientist, Hanson Robotics)

Goertzel goes on to point out that by his definition, "an awful lot of things not naturally considered intelligent may be viewed as intelligent to a limited degree." Here, we're getting closer to a fully inclusive definition, but it feels like we're still approaching the question in binary terms when it's becoming increasingly evident from this survey that most aspects of intelligence may in fact exist on a spectrum.

In building an exhaustive list of intelligence definitions, AI researchers Shane Legg and Marcus Hutter sought out as many distinct voices as possible.[6] (Their document is often cited as the most thorough list of such definitions.) Nevertheless, they found these characterizations coming up short and so included their own broad definition of intelligence. They did this with an eye to developing a universal intelligence test in order to better identify and measure intelligence, regardless of where it is encountered or what form it might take.

> "Intelligence measures an agent's ability to achieve goals in a wide range of environments." (Shane Legg, machine learning researcher; Marcus Hutter, AI computer scientist)

Here, Legg and Hutter use the term *agent* to specify an entity that may be biological, technological, alien, or otherwise. They and other researchers have focused in recent years on designing methods for measuring any form of intelligence, regardless of level, type, or substrate (the underlying structural basis on which something is built: proteins, neurons, silicon microchips, etc.). Ideally, a universal intelligence test would be able to adapt to any agent, presenting challenges that are appropriate not only to the level of intelligence, but also to the agent's form, structure, substrate, and capabilities.[7] Such an approach recognizes that while machines may not yet be at the stage of true intelligence, that day is rapidly approaching. Being able to measure and assess intelligence levels and milestones throughout the process will provide considerable benefit and data.

Nevertheless, Legg and Hutter's definition of intelligence, while certainly broadened, still continues to make assumptions that are very human in nature. "Achieving goals" can be broadly interpreted, but still carries specifically human undertones.

Famed physicist Stephen Hawking takes a still more general stance.

"Intelligence is the ability to adapt to change." (Stephen Hawking, theoretical physicist, cosmologist)

This feels like we're making progress toward eliminating human-centric biases and expectations of value and behavior. There's a single key condition that would unify a very broad range of intelligences. But is this too broad a definition? Have we opened the door too far? After all, a mercury thermometer adapts to changes in temperature, yet we can hardly think of it as having intelligence.

Before we begin closing this newly opened door, however, let us consider some recent trends in Big History, which may provide a different way of looking at things. Big History is an academic discipline that sets humanity's story within the story of the universe, framing history within vast timescales, often beginning with the Big Bang, in order to explore and discover universal patterns. According to the founder of the Big History Project, David Christian, "What Big History can do is show us the nature of our complexity and fragility and the dangers that face us, but it can also show us our power with collective learning." The perspective afforded by Big History can offer not only new insights into the world and universe we live in but also, as we will see, hints at potential solutions for many of the challenges we face.

Based on some ideas in this multidisciplinary school of thought, many of nature's processes may be far more intricately connected than we've traditionally considered. Cosmological and stellar evolution may share features, relationships, and possibly even underlying processes with planetary development and the origins and evolution of life, along with human, social, and technological development. Moreover, a trend toward increasing organization and complexity appears to exist that belies assumptions about the uncaring nature of the universe.

Why should such different scales of the cosmos organize themselves in such interconnected ways? Perhaps more importantly, why should it organize itself at all? What don't we understand about nature that would allow it to progress steadily in a direction that seemingly flouts the second law of thermodynamics, which states that the total entropy of an isolated system can never decrease over time?

Of course, none of this truly defies thermodynamics. But given the nature of entropy—the tendency for everything to run down and lose structure over time—it seems like something is working against this process. It is almost as if a leaf were drifting on a flowing river, except that through a series of random movements it continually, consistently moved upstream toward the headwaters. Needless to say, this is far from the behavior we would expect to see.

We'll discuss entropy in the coming chapters, but for the moment suffice to say it is the inescapable inclination for all things, including the universe itself, to run down and become more disordered. This decline will occur in all isolated systems that receive no additional energy from external sources. This inevitability of nature is formalized as the second law of thermodynamics and is why perpetual motion machines are impossible, regardless of how clever and convincing the design.

However, there is an outgrowth of this law that at first glance seems counterintuitive: the localized trend toward increased complexity. According to a concept known as *causal entropic forcing*, the development of emergent phenomena can be driven by thermodynamics and can lead to pockets of greater complexity. This process results in a corresponding reduction of local entropy, while increasing overall universal entropy. It should be emphasized that by definition these pockets of emergent organization are not isolated and use external sources of energy. Therefore, the laws of thermodynamics remain unbroken.

Physicist and mathematician Alexander Wissner-Gross has developed a number of mathematical computer simulations that demonstrate the idea of causal entropic forcing[8] applied to evolving computer simulations. These suggest a view of nature that is very different from our more classical interpretations. Based on this work, Wissner-Gross has developed an equation for intelligence: $F = T \nabla S\tau$. Stating his equation succinctly in lay-speak, Wissner-Gross defines intelligence as:

"Intelligence acts so as to maximize future freedom of action."
(Alexander Wissner-Gross, physicist, mathematician)

"Future freedom of action" is a daunting phrase, not least because of another consequence of thermodynamics and entropy. Time travel is for all practical purposes a one-way trip, making actual travel into the past an impossibility, at least based on our current understanding of physics. Therefore, knowledge of whether something increases "future freedom of action" should be impossible to obtain. However, if a number of variations should attempt to occupy a particular opportunity space, we could find, in retrospect, that one was the best fit for maximizing future freedom of action—without needing to travel back in time to deliver the news. At one level, it is analogous (or perhaps even related) to the concept of evolution, which incrementally but blindly takes us from one structure or species to another without deliberate intent.

All of the definitions given in this chapter serve a purpose in their own right, whether exceedingly broad or highly restrictive. But in the case of this last definition, we suddenly find ourselves occupying a universe filled with all manner of potential intelligences. Perhaps more accurately, time and again, the universe has seen different features and aspects optimizing in this way, leading to new states of increased complexity. Here in our humble corner of the universe, 13.8 billion years from when it all began, cognition and consciousness have been the resulting outcome, but after this, who knows what form such optimizations might take?

Though it should be unnecessary, let me state unequivocally that protons, stars, and RNA, in and of themselves, do not *think*. But perhaps in trying to define intelligence, we have been focused on the wrong thing; not seeing the forest for the trees, as it were. If we shift our perspective and instead view intelligence not solely as cognition but as a manifestation of much larger optimization phenomena, we may end up seeing the universe in a far different light.

From this perspective, different forces, processes, life-forms, and intelligences will manifest at various stages of transformation, given a sufficiently open environment or domain. In a domain in a state of high entropy, with enough time and energy differentials to drive a vast number of informal experiments, one or more new forms (of energy, matter, life, intelligence) may have the opportunity to manifest and be perpetuated. Such a process could explain how the universe established the perfect conditions for the existence of life, eventually giving rise to intelligence, consciousness, and whatever states of increased complexity and maximized freedom of action may manifest in the future.

The processes that gave rise to life and intelligence connect us, in a very direct way, all the way back to the origins of the universe. But to consider this idea more thoroughly, we'll need to go back to where it all began. Fortunately for us, the spaceship of my youth has undergone significant upgrades in recent decades, allowing us to journey all the way back to the earliest moments of our universe.

CHAPTER 2

THE ORIGIN OF EVERYTHING

"Nothing can create something all the time due to the laws of quantum mechanics, and it's fascinatingly interesting."

—Lawrence M. Krauss, theoretical physicist,
cosmologist, author

It seems reasonable to suggest that any exploration of the origins of intelligence must begin from the very beginning. The Alpha moment. When the laws of the universe were first and firmly established by the cosmos, leading to all the forces and interactions that would eventually give rise to the universe we know and see today. While all of this may seem a long way from our traditional concepts about intelligence, these initial chapters explore certain basic processes of the universe that appear to ensure localized trends toward ever increasing complexity that eventually emerges as intelligence.

This is why we find ourselves in the deep past, 13.799 billion years ago, nestled in the trusty spaceship of my youth (significantly upgraded since its early days), suspended at the very beginning of time (at least as this universe knows it), on the cusp between space and not-space, between nothingness and all the energy and matter that are and ever will be.

It's zero hour and the universe doesn't truly exist yet. For that matter, we don't really know how everything gets started. It may be that this is a time/place/instant of near infinite heat and density, its origin a mystery.

Or at least that's one possible version. Once it was discovered in the early twentieth century that the universe is expanding, the implication was that if you trace that expansion back far enough in time, you reach a beginning when it would have been infinitely dense and infinitely hot. But if this was so, then where did this superhot, supermassive singularity come from?

Though this gravitational singularity would have had some similarities to those other singularities known as black holes, it is difficult to say just how similar. For instance, each may have been nearly infinitely dense, creating conditions so extreme it challenges our understanding of many of the laws of physics. But whereas a black hole exists in space, for the singularity of the Big Bang, space didn't really exist at all. There was no outside as we understand it, and so there are no reference points. Therefore, while we can theorize, we can't say with certainty how this infinitely dense point might have come about.

Another, more recent theory about the origin of the Big Bang is that it all began from *nothing*. While this can be a challenging idea to accept, growing evidence suggests this may actually be what happened. When most of us think of nothing, we think of an absence of everything. But physicists will tell you that *nothing* is actually far more complex and interesting than we give it credit for.[1] Because of general relativity and quantum mechanics, including the effects of Heisenberg's uncertainty principle, something truly can arise from nothing.[2]

Physicists talk about nothing in a number of ways, including the idea of a *metastable false vacuum*.[3] This refers to empty space where virtual quantum particles continually pop in and out of existence, faster than we can measure. This sounds bizarre, yet it is completely acceptable within the laws of quantum physics. The thinking is that at some point, some of these virtual particles extended the false vacuum beyond a certain critical threshold, leading to a runaway expansion. With that, our universe was born.

However the Big Bang got started, this marks the beginning of the Planck epoch, the earliest stage of the universe when random quantum forces dominated over every other fundamental force, *including gravity*. Yet despite its enlarging to many times its original size during this period, we can see no evidence of this from our ship, which for the moment still exists apart from the singularity universe described in this scenario. This is because there is nothing for the universe to expand into. There is no external frame of reference, and so it is impossible to gauge its growth or any other changes. Therefore, to realign our point of reference, we'll rescale our ship in time and space, teleporting from not-space to inside this miniscule but rapidly expanding cauldron, so we can better observe the changes that are taking place. Given how incredibly tiny the universe is at this moment, our ship has been reduced to only a fraction of a Planck length, or 10^{-35} meters.

For those unfamiliar with that notation, there will be frequent use of exponents in this chapter to prevent the pages from being filled with lots and lots of zeroes. For example, a Plank length of 10^{-35} meters can also be represented as 1/100,000,000,000,000,000,000,000,000, 000,000,000th of a meter. (The negative sign transforms the number into its reciprocal.) The estimated temperature of 10^{32} kelvins at the end of the Planck epoch can also be written as 100,000,000,000,000, 000,000,000,000,000,000 K. (For comparison, the surface of our own sun is about 5,778 K and its center is approximately 15,700,000 K.)

As mentioned, our ship has rescaled *time* for us as well. This is critical, because so much is about to happen in such an infinitesimal span of time. A tremendous amount of the universe's future will be established in the barest fraction of the first second. It is estimated that the Planck epoch lasted until 10^{-43} seconds after the initial moment of expansion. (For comparison, the fastest laser pulse ever created in a lab is about 2×10^{-21} seconds, so that earliest epoch occurred about a billion trillion times faster than that.)

We have now reached the point when the singularity had grown to exactly one Planck length. Until this moment, it is assumed that quantum gravitational effects dominated throughout the then infinitesimally tiny universe. There remains some disagreement on this early history, however, because in modeling the earliest moments, general relativity breaks down, making it impossible to do much more than speculate on the initial properties of the universe. As we'll see, as the universe expands and cools, the fundamental forces and physics of the classical world will begin to emerge.

Not surprisingly, the end of the Planck epoch marked the beginning of that transition. It's generally assumed that the Planck epoch ended with gravity breaking away from the electronuclear force (the strong, weak, and electromagnetic forces remaining unified in that force). The subsequent grand unification epoch would last until 10^{-36} seconds, when the electronuclear force separated into the strong and the electroweak forces. This led to the universe being filled with a quark-gluon plasma that would in short time give rise to all of the subatomic particles and fields of our present-day universe. As the universe cooled to a mere 10^{28} kelvins, the electroweak epoch began, lasting until 10^{-32} seconds as the weak force and electromagnetism finally separated into the two remaining classical fundamental forces. Gravity, the strong force, the weak force, and the electromagnetic force would now exist for the lifetime of the universe.

Because there remains considerable speculation about the earliest stages and processes of the Big Bang, we may never completely understand the origin of our universe, not even its size and temperature in those earliest instants. But what is all but definite is that throughout the early epochs that set the stage for the universe as we know it, many important and extremely powerful transformations took place—transformations critical to the eventual development of life, intelligence, and nearly everything else.

During those first moments, numerous phase transitions occurred as the universe expanded and subsequently cooled. This may or may not have included a period known as the inflationary epoch. While the Big Bang theory explains much of our early universe extremely well, and many predictions about it have been proven beyond a reasonable doubt, inflation theory, while promising, is somewhat more hypothetical. According to the theory of cosmic inflation, a force called an inflaton field caused the universe to grow far faster than the speed of light, expanding to become 10^{26} times larger (perhaps even more) in a fraction of an instant.[4] From its initial state when the universe was less than one Planck length across (10^{-35} meters), it swelled to perhaps the size of a grain of sand. As a comparison, this is about the same relationship we would see if we suddenly lengthened a centimeter-long piece of string so that it instantaneously could reach the edge of the modern-day universe.

But wait! you may be thinking. Nothing is supposed to be able to exceed the speed of light. While this is true for matter and energy *in* the universe, this is a special case because these changes were occurring to *space itself*. It feels like a cheat, but it actually makes sense, since special relativity is about actions and measurements taking place between objects within a common reference frame of space-time. But relativity doesn't prohibit space from expanding far faster than light speed.

The inflationary epoch would have been essential to everything that came after because of how it altered our universe. Starting off from a highly curved geometry, the shape of the universe would have been flattened out by inflation. Additionally, as the inflaton field caused the cosmos to rapidly expand, minor differences of density from quantum fluctuations led to slight variations in the distribution of matter everywhere. These "seeds" allowed gravity to concentrate more and more matter, eventually leading to the formation of the galaxies and stars hundreds of millions of years later. But before that could happen, the cosmos would have to cool and change again and again.

Many transformations occurred in the seconds, minutes, and millennia that followed. About ten seconds after the Big Bang, the universe had cooled to a mere billion degrees, the temperature at which atomic nuclei could form. For the next seventeen minutes, a process known as the Big Bang nucleosynthesis formed all of the stable hydrogen and helium nuclei in the universe. (A relatively small amount of lithium was also created.) Then, as the universe cooled to approximately ten million kelvins, the fusion process shut down. Still, it was far too hot for the nuclei to bind to free electrons in order to form neutral atoms. It would be hundreds of thousands of years before these could form.

Directing our ship to leap forward in time past most of the photon epoch, the era when the universe was filled with a plasma of photons, hydrogen, and helium nuclei and electrons, we find ourselves nearly four thousand centuries after the Big Bang. Since the first fractions of an instant following the birth of the cosmos, the universe has been expanding. As it has done so, its density has diminished and its temperature has cooled. Starting from a singularity with temperatures estimated to be well over 10^{32} kelvins (that's 100,000,000,000, 000,000,000,000,000,000,000 K), 379,000 years later the universe has expanded to approximately forty-two million light-years across, with a temperature in the vicinity of 4,000 K. (One light year is just a little less than six trillion miles.) At this point, the epoch known as *recombination* begins. The universe has cooled sufficiently so that photons have become much less energetic, allowing free negatively charged electrons to be captured by the ionized nuclei, binding to the positively charged protons. Throughout the universe, these subatomic particles combine to form stable neutral atoms, the hydrogen and helium atoms we interact with today.

Up until this time, the universe was filled with a plasma of ionized hydrogen and helium, free electrons and photons, which continually interfered with the movement of photons. This interference resulted in a very short *mean free path* for all the photons in the universe, so that it would have been impossible for electromagnetic radiation to travel in a straight line for any significant distance. The forming of stable neutral atoms has the added effect of "decoupling" the photons from matter, allowing them to travel freely through the cosmos. By the end of this decoupling stage, nearly all of the photons making up all the electromagnetic radiation in the cosmos, including visible light, could move about freely. The universe, which has up to now been opaque, quickly becomes transparent.

The end of the recombination epoch is as far back as we can directly observe the physical universe because of this critical phase state change, a shift in the universe's physical properties. The transition is demarcated by a barrier of ionization that we now refer to as the cosmic microwave background, or CMB. At the time it originated, it could have actually been observed as visible light and infrared radiation. From our ship, situated when the universe was only 379,000 years old, we see the CMB as an orange glow. However, in our present day, the light has traveled 13.799 billion years to reach us, and it has done it through rapidly expanding space. Because of this, the ancient light of the CMB has been "redshifted,"[5] or stretched, as described by special relativity, so that now, in the twenty-first century, we don't see it as visible light at all. Instead, our telescopes receive it as radio waves—microwaves, to be more specific.

What does all of this have to do with intelligence? Physicists have observed for decades that if the many laws and properties of the universe had differed by even a miniscule amount, the stars, the planets, and life in the cosmos would not have been possible. How did we get so lucky that the universe formed in exactly the right way for us to be here?

A set of philosophical considerations known as the *anthropic principles* states that the universe has to be compatible with the conscious life that exists in and observes it. While this may seem self-evident, the anthropic principles could potentially act as a selection mechanism, instead of an astronomically unlikely roll of the dice. Assuming the concept of a multiverse is valid—that countless universes exist beyond our own—then we exist and are able to observe the universe because it is one of probably many universes in which the conditions are just right for our existence. In a way, our "Goldilocks" universe is the outcome of a process by which a population of universes maximized future options. While the multiverse concept is very speculative, many leading physicists are proponents of it, including Max Tegmark, Alan Guth, Brian Greene, Andrei Linde, Michio Kaku, and Neil deGrasse Tyson. While this concept may or may not ever be falsifiable (the potential in principle to prove something false, a hallmark of scientific inquiry), that doesn't mean it isn't possible, only that it may not be provable.

This brief recap of the Big Bang barely scratches the surface of everything that occurred in establishing the early universe. But despite our having skimmed over many major events that followed the Big Bang, the stage is now set for the next crucial juncture that will eventually make intelligence in the universe possible.

Creating Chemistry

From the safety of our ship, it is evident there is not much to see in the nascent cosmos just yet. So far everything that's transpired has been at the subatomic scale, if not far smaller than that. It's approximately 150 million years after the Big Bang, and the expanding universe has cooled to a chill 60 K. If nitrogen existed, which it doesn't yet, it would be frozen solid. Nearly everything in the universe has dropped˜ to this temperature, so of course we should probably discuss those parts that haven't.

One long-term effect of the inflationary epoch was to take what had been a nearly uniform canvas of space and exaggerate the very minor variances left over by the quantum fluctuations of the early universe. This allowed pockets of matter to begin to accumulate, creating regions of gravity in the relatively empty vacuum of space. Without this, it's unlikely there would have been enough substantial differences in density to have been able to slowly, incrementally draw enough material together to produce stars.

But there was, and they did, and here we are. Nearly every aspect of the universe plays a role in the evolution of matter, life, and intelligence. But it is the stars, the solar furnaces, that were critical to every other process that would follow, because this is where all of nature's elements were created.

The famed astrophysicist Carl Sagan perhaps expressed it best when he said: "The nitrogen in our DNA, the calcium in our teeth, the iron in our blood, the carbon in our apple pies were made in the interiors of collapsed stars. We are made of star stuff."

So, when gravity began drawing material together, it would become a very big deal, both figuratively and literally. Little by little, the minute gravitational forces between those atoms drew them together into thin filaments of gas, filaments that would eventually aggregate into vast clouds of hydrogen and helium, virtually the only elements to exist at that time.[6] Over hundreds of millions of years, these gravitationally bound structures would merge to become proto-galaxies, then galaxies, and eventually galaxy clusters. As this happened, the coalescing of the gas incrementally raised its density, leading to increased activity between the molecules, which in turn raised its temperature in those regions where it was most concentrated. Then the hydrogen atoms merged to create hydrogen *molecules*, which cooled the hot, dense gas, reducing its internal pressure and allowing it to collapse still further.

In addition, the modest movement of the incoming gas slowly caused each region of increasing density to spin around its gravitational center. As gravity continued to draw it inward, the law of conservation of angular momentum dictated that this material would spin faster, for exactly the same reason a spinning ice skater turns faster as she draws her limbs close to her body. As more cosmic material was drawn in and it spun still faster, this gas flattened into an accretion disk and its center incrementally became more compact. Once there was enough material, the center of the gas disk collapsed under its own mass in a runaway reaction, the gas ball growing so dense at its center that it began to fuse the atoms in its inner core. This center was destined to become a new star, and the rest of the accretion disk became the planets, the moons, and the asteroids of that new solar system.

The comparatively warmer conditions of the early universe meant that far more gas had to accumulate before it acquired sufficient mass to collapse and form a star. As a result, many of these first-generation stars were 500 to 1,000 times more massive than our own sun. Such a massive star may have had a surface temperature in the range of 100,000 K, nearly eighteen times that of the sun. As a result, most of the massive star's energy would have been radiated in the ultraviolet range (though there still would have been plenty of energy remaining in the visible part of the spectrum—a human could have seen it). It was in that moment of first stellar fusion that the cosmos experienced light as it never had before, as it witnessed the birth of its very first star.

We don't know anything about that first star—not its size, temperature, or life span. It is long gone and unlikely to ever be identified. What we do know is that it was not unique. It was soon followed by thousands, then millions, and later billions of stars. Once the right conditions were in place, there was no stopping the process.

The stars in this first generation—often called Population III stars—were very different from those that would follow. They were formed entirely from the primordial hydrogen and helium that was generated in the Big Bang nucleosynthesis that took place in the first minutes after the universe was born. While later generations of stars would contain a range of elements, during this earliest era, there were no other atoms to gather and fuse other than hydrogen and helium. (An estimated 1 percent of primordial matter in the universe may have been lithium, the third element.)

Because of their tremendous size, these first-generation stars quickly used up their fuel, incrementally fusing hydrogen and helium into metals like lithium, beryllium, carbon, nitrogen, oxygen, and so on up through iron. This set the stage for the chemical needs of the universe to come.

This process of stellar nucleosynthesis is related to but extremely different from the Big Bang nucleosynthesis that took place in the first minutes of the universe. That first event formed nearly all of the hydrogen and helium nuclei in the universe. These charged nuclei later combined with electrons to form neutral hydrogen and helium atoms during the *recombination*.

However, stellar nucleosynthesis is very different, in that those hydrogen and helium atoms now undergo fusion, producing heavier elements as well as releasing a substantial amount of energy in the process. In the fields of astronomy and cosmology, all of the elements other than hydrogen and helium are referred to as metals. *Metallicity* refers to how rich or poor a star is in these metals (and, since they probably formed from the same accretion disk, any planets surrounding that star as well).

Though there are many ways a star's life can end, usually based on its size and metallicity, one of the universe's most powerful cataclysmic events occurs when a star explodes in a supernova. Because of their size, this should have been far more common in those earliest, massive stars. A supernova generates a combination of enormous pressures and high temperatures not found anywhere else in the modern universe and is responsible for the formation of those elements in our universe that are heavier than iron. A supernova can also leave behind a core that could collapse into a neutron star or a black hole. Perhaps most importantly, while the mass and makeup of each star determines how it will die and what elements it will produce, nearly all stars contribute to altering the chemical makeup of the universe.

Today we see a cosmos filled with older, metal-poor Population II stars and younger, metal-rich Population I stars. This latter group probably offers the best possible conditions in which life can potentially form. Could life and intelligence have evolved around an early Population III–star solar system? It seems unlikely, given how little time was available for life to take hold. But far more importantly, the absence of heavier elements would have made all but the most basic chemical reactions impossible.

However, in the generations of stars that would follow, this wasn't the case. Over time, an expanding, cooling universe and the increasing prevalence of heavier metals made extremely massive stars far less common. This led to stars and solar systems with much longer lifetimes. Along with this, more and more heavy elements were being fused with each new generation of solar furnaces, either as products of the fusion performed throughout their lives or in the cataclysm of their death throes.

Both of these conditions would vastly increase the likelihood of generating biological life and intelligence. Based on our experience on Earth, the ten-million-year life span of the first stars was a drop in the evolutionary bucket. (It is barely the amount of time that separates *Homo sapiens* from the last common ancestor we share with chimpanzees.) But if a star is of a more moderate size and therefore survives something on the order of ten billion years, there is obviously more than enough time for simple and complex life to take hold (our own sun being 4.6 billion years old).

As we've seen throughout the early life of the expanding universe, a wide range of events took place as the cosmos cooled, often leading to phase state changes that altered the universe dramatically. Many of these occurred at quantum scales and energy levels that we are still far from fully understanding: the separating of the fundamental forces; the imbalance in the quantities of particles and antiparticles; the decoupling—the deionization of the universe. But one event we understand reasonably well is how variations in the distribution of matter and energy would produce even greater variations as the universe expanded, leading to differences in *energy potential* between regions. This had an enormous impact on the universe nearly from its beginning and will continue to do so until its very end.

Whether those initial variations came about because of minute variations from quantum fluctuations that were later amplified during the inflationary epoch and the hyper relativistic expansion of the universe, or because of some other cause, the point is it happened, and it happened in a very particular way. A very basic process used energy—defined as differences in potentials for work between entities or regions—to increase complexity throughout the universe. From a nearly uniform plasma, minor differences created a foamlike structure of filaments and voids that would in time become unimaginably massive. Without the gravitational differences this led to between these different regions, the stars would never have formed. In which case the cosmos would've been a dark, permanently empty place.

This is a theme that will be repeated throughout this book. As we see time and again, during the development and evolution of life, as well as in the rise of intelligence and culture, a similar pattern occurs. The universe organizes at every scale in ways that seem to lead to local organizations of greater complexity while increasing entropy for the universe overall. This trend continues on a trajectory of escalating complexity, resulting in systems that at some point acquire sufficient volition that we begin to call it intelligence. Recall our last definition from chapter 1: "Intelligence acts so as to maximize future freedom of action."

If such actions occur again and again, a pattern emerges. Is it a different way of looking at intelligence? Or viewed from a different perspective, is what we've traditionally called "intelligence" merely a point or a region on a far larger spectrum, one describing systems that attempt to optimize in order to better perpetuate themselves in the universe?

We will be exploring such moments of fundamental transition throughout this book. With each step, be it gravitational, chemical, biological, cultural, or technological, something new occurs. Each of these transitions establishes a change, after which nothing will ever be the same again. We can point at different explanations, but the best and simplest commonality is one of emergence and complexity. By its nature, emergence is both unpredictable and a game changer. As we will see, emergence materializes out of complexity, transcending the sum of its parts in ways that are both synergistic and unexpected. We can't anticipate how, or usually even when, something will arise in this sense, but when it does, it will be like nothing that's come before it.

The universe is full of these moments, and in better understanding these transitions, these inflection points, we may be better preparing ourselves to weather future moments of such transformational change.

In looking back along such a spectrum, if we find ourselves questioning the labeling of certain processes and agents as intelligent, that may be our bias showing. However, if we look ahead at the next major milestones according to this definition, at systems of far greater complexity and having emergent properties we don't fully understand, will we recognize those as intelligence either? Perhaps it is time we looked this question from a somewhat different perspective.

CHAPTER 3

ENDURING ENTROPY

"Entropy drives both the increase and decay of complexity."
—Matt O'Dowd, astrophysicist, professor,
host of PBS *Space Time*

It is a hundred million years after the Big Bang, and a tenuous thread of hydrogen molecules drifts effortlessly through the cosmic vacuum, stretching out across the light-years. It moves almost imperceptibly in a vaguely general direction, beckoned by an unseen summoner. Millennia pass, and after covering vast distances, the molecules now move with a little more haste. The pattern continues until the tendrils of gas slowly converge on a nexus, a meeting point that has been incrementally accumulating material from every direction. Drawing together slowly, though considerably more rapidly than before, the gas gradually accretes into a boundless cloud, an anomaly in this vast near-nothingness. Though far less concentrated than the air we breathe, it is already becoming the densest structure in the universe. Time and more molecules will lead to still greater densities, new structures, and the beginning of processes the universe has never seen before. And this is only the beginning. In time this assemblage will become a solar nursery from which the first stars will be born. It's all a matter of time, gravity, and *entropy*.

Slowly, imperceptibly at first, this aggregation of gas, this vast cloud begins to spin.

The universe is vast beyond imagination. At the same time, its power is just as incomprehensibly immense, a colossus of potential driven by its tremendous scale and structure. We've talked a little about how this came about, but perhaps even more importantly we should ask: Why did any of this happen at all? Why do the immense celestial objects we call stars, galaxies, nebulae, and black holes even exist?

In the earliest moments of the Big Bang, the universe was nearly uniform. As we saw in the previous chapter, quantum fluctuations, of the same type that possibly set off the Big Bang in the first place, also prevented the universe from being completely smooth. During the universe's expansion, particularly during the inflationary epoch, these variations in density grew far more pronounced, resulting in areas of greater and lesser density. This allowed gravity to be stronger in some regions than in others, making it possible to draw material together, further concentrating hydrogen, helium, and lithium atoms into pockets of space that would eventually become the first stars.

These initial variations in density were critical, because in time they led to still greater variations in the extreme. It is estimated that in our present-day universe, the typical vacuum of space has roughly one atom per four cubic meters. Yet our own sun has a density that is roughly 3×10^{30} times greater than that, and a neutron star is 4×10^{14} times greater still. (A typical neutron star has around 12×10^{44} atoms per cubic meter. A teaspoon of this material would weigh around ten million tons!) This is a degree of difference that should give us pause, even as it astounds us. Because differences in energy, be they gravitational, electromagnetic, or chemical, are what make the universe go around, so to speak.

Whenever there is a difference between two systems, there is the potential for change, or "work," as defined by thermodynamics.

This work is the energy transferred by a system to its surroundings, accounting for all forces exerted by the system. We see this all the time in our everyday world. The ocean tides occur because of differences in gravity caused by the moon, the sun, and variations in Earth's surface. Lightning is the result of a buildup of positively and negatively charged particles in clouds. When this differential exceeds a certain threshold, the particles discharge, either within the cloud or between the cloud and the ground, equalizing the system. A hot cup of coffee cools because of the imbalance between its temperature and that of its surroundings. But why do these systems change at all, transforming themselves from one state into a less energetic one? Why don't they simply remain as they are?

We can be thankful that they don't, because these forces of change are what led to stellar evolution, chemistry, the formation of the planets, and the origins of life (and that's just for starters). But to answer the question, they all owe their existence to *entropy*.

Entropy is the great unsung hero of time and space, an oft-misunderstood aspect of nature that is an emergent function of probability and statistical mechanics. Entropy is the unrelenting process by which particles, energy, and systems progress from a less probable distribution toward more probable ones. As a measure of the order in a system, we can generally say that the greater the entropy, the greater the disorder. For example, if you have a container filled with gas molecules, and all of those molecules are arranged on one side of the container, that is a low probability distribution and so we say its entropy is low. There is far more order in that system than if it were merely a random configuration. But give it a few moments and let the molecules move and redistribute throughout the container into a more even distribution, and the entropy of the system is said to be much higher. As the system reaches equilibrium, the distribution is at its most probable, statistically speaking, and its entropy is at its maximum. This is the basis of the second law of thermodynamics, and it is why *everything* runs down and decays. By contrast, highly complex systems such as paramecium, supercomputers, and human beings are extremely improbable and consequently are considered low entropy systems. Because of this, as their entropy increases, they fall apart, stop working, and eventually cease to exist. From a purely statistical standpoint, any closed system will follow this course—unless it receives outside energy, in which case it's no longer a closed system.

Entropy is so often envisioned as the bad guy, the antagonist, the bringer of end times. All of that is actually backward. Entropy is the force that's not a force. It is the power without a throne, so to speak. The universe functions, flows, shifts, and changes because entropy enforces change, purely because that is the nature of probability. Without entropy, the planets wouldn't have formed, water wouldn't flow, and the sun wouldn't shine. Without entropy, the concept of free or usable energy would be meaningless, because energy couldn't change or bring about equilibrium in a system or have the potential for performing work. So yes, people die and stars burn out and the universe will end someday, all because of entropy. But without it, none of them would have been able to exist in the first place.

It's funny to think of probability being such a tremendous force of nature and influence on the universe. So often we think of it as a benign artifact of weather forecasts, research studies, and roulette wheels. But imagine a world in which the laws of probability didn't hold, where someone might flip a coin a million times and never have it come up tails. Where airliners, built to the highest standards and specifications, nevertheless fell out of the sky every time they flew. Where a common element like carbon could unexpectedly decay at any moment, releasing a deadly storm of neutrons. Our world and the universe work because the laws of probability remain constant and immutable.

Probability is used to study the properties of complex systems having many degrees of freedom in a field known as *statistical mechanics*, which was initially developed in the 1870s by Austrian physicist Ludwig Boltzmann. Statistical mechanics is essentially applied probability as it relates to large numbers of particles, particularly their energy distribution, and it tells us why large-scale systems have the properties they do. Statistical mechanics describes a system's many microstates and their resulting macrostate. A microstate statistically defines each particle in a system—its position, temperature, spin, etc.—whereas a macrostate defines the system's overall properties, such as its temperature, pressure, and form. A large number of microstates can potentially result in the same macrostate, even though each microstate is unique. For instance, the water molecules in an ice cube are in a relatively limited number of microstates, so the cube is said to have lower entropy compared to when the exact same molecules have been heated and turned into vapor. There are many more possible configurations for those gas molecules at any given moment, so the greater the number of microstates that make up its macrostate, the greater the system's entropy. This form of thermodynamic entropy is often called Boltzmann entropy and is calculated using that physicist's now-famous formula: $S = k \log W$.

Skip ahead three quarters of a century and Claude Shannon, the father of information theory, has developed a measure of how much information can be contained in a message. One of Shannon's great insights was that the amount of information in a given message is directly related to its uncertainty. That is, the greater the average uncertainty in a message, the more information it contains. (Whether or not that information means something to us is another matter.) Shannon called this uncertainty "entropy" (also known as information entropy or Shannon entropy). For instance, a source that always produces the same output, such as a string of Zs, is entirely predictable and its entropy is effectively zero. This string can be highly compressed and provides no surprise or information to the receiver. On the other hand, a source that generates bits randomly, each with an equal likelihood of zero or one has maximum entropy. Any information from this latter system has the greatest average uncertainty and carries the greatest surprise and therefore information in its message.

Many people were struck by how similar Boltzmann's and Shannon's formulas were. This has led to many decades of disagreement over whether or not the similarity is anything but superficial. Many people have pointed to the obvious differences between physical and informational systems. While there will probably always be argument over this, there is a growing consensus that the two are in fact closely linked.

Information does not exist in isolation in the universe. It is always a feature of and embodied by physical systems, whether these are photons, atoms, or cells. Change just one property of that object or system, even in the slightest degree, and you've not only altered the information contained within it, you've also increased the entropy of the universe. According to what is known as the Landauer limit, altering even one bit of information in a system (the smallest unit of information as defined by Shannon) has been shown to require energy and therefore must be accompanied by a corresponding increase in entropy.[1]

Likewise, every physical object and system in the universe, from the smallest quantum particle to the most massive quasar, exists because of, and is defined by, the information it embodies. So, while there are certainly differences between these two forms and measures of entropy, in some respects it might be better said that thermodynamic entropy is a subset of information entropy. One way to think about this is that the information that defines a system also defines the properties of its thermodynamics—its microstates and macrostate. Given this relationship, throughout this book, the term *entropy* will be used generally, unless there are specific reasons for denoting otherwise.

When the Big Bang commenced, the universe was at the lowest entropy it would ever have. From that moment forward, the entire cosmos has been expanding, cooling, and running down. It will continue to do so until everything—and that means *everything*—reaches equilibrium, in what has been called the heat death of the universe. But lest you feel anxious about this, consider that while the universe is currently 13.799×10^9 years old, the end of everything is calculated to arrive no sooner than 10^{100} years from now, and by many measures it may well be much further off than that!

But if the universe is always running down, dissociating, becoming less and less ordered, how can complex things like plants, paramecium, and people even exist? It's a big question. The answer is that, while entropy is in the continual process of degrading useable energy into waste heat and noise, it is also the primary driver of complexity. This happens, at least in part, because these complex systems, including ourselves, increase the speed and efficiency at which the universe processes and dissipates energy.

Does that seem counterintuitive? After all, here we are bringing order to our little pocket of this cold, dark, chaotic universe. We build cathedrals, write sonnets, perform symphonies. We created a globe-spanning technological society. How can we be contributing to entropy?

In our own infinitesimal way, our very existence causes the entropy of the universe to increase. We're not unique in this. Every system that acquires, stores, processes, or expresses information and energy, be it a star, a chemical reaction, a human being, or technology, uses that energy to reduce its local entropy in order to establish and maintain its complexity. But because of the laws of thermodynamics, this local reduction of entropy has to be balanced out by increasing entropy in the rest of the universe. Generally speaking, the greater the complexity of the entity or organism, the greater its energy use (after factoring for scale) and the greater its overall production of entropy.

Renowned Harvard astrophysics professor Eric Chaisson has explored this growth in complexity and energy processing in a number of papers and books on cosmic evolution, notably using the concept of "energy rate density" to trace certain patterns. According to this work, though galaxies and stars are far more powerful than almost any other system in the universe, their energy flow over time relative to their vast size is far lower than that of plants, animals, and human beings. In fact, each of the latter is exponentially greater by this measure, even as each has evolved and appeared on far shorter and more recent timescales as we approach the present day. In his 2010 paper, "Energy Rate Density as a Complexity Metric and Evolutionary Driver,"[2] Chaisson describes it this way:

What seems inherently attractive is energy flow as a universal process—specifically, energy rate density as a single, unambiguous, quantitative measure of complexity—that helped to control entropy within increasingly ordered, localized systems evolving amidst increasingly disordered, wider environments, indeed that arguably governed the emergence and maturity of our galaxy, our Sun, our Earth, and ourselves.

Why should this be? How can order spontaneously arise out of chaos? Doesn't this run counter to the very laws of thermodynamics? The last question is the easier one to answer, and the answer is "No." The irreversible nature of entropy and the second law of thermodynamics apply to isolated systems; however, if additional energy can enter a system, whether as differences of gravity, solar energy, undersea geothermal vents, or any of a number of other possible external sources, certain aspects of that system may have an opportunity to use that energy to increase in complexity, thereby reducing its local entropy. Within the laws of thermodynamics, this is totally acceptable.

Modeling "energy flow in complex systems from the Big Bang to humankind," Chaisson finds that as we progress through the evolution of the universe, increasingly complex systems appear, using energy at consistently faster rates. From galaxies to stars to planets to plants to animals to human society, each processes energy more rapidly and efficiently than systems of lower complexity. This makes sense in terms of the relationship between complexity and entropy. A planet is far more complex than a star, just as an animal is generally much more complex than a plant. Without a regular influx of energy to offset the effects of entropy, a more complex entity will fall apart and cease to function relatively rapidly. Therefore, gram for gram, the more complex something is, the more energy it is going to need in order to establish and maintain that complexity.

Measuring energy rate density in terms of ergs per second per gram, Chaisson finds that human society processes energy 250,000 times more rapidly than our own sun. At first glance this seems absurd, given the immense power of a star, but when we factor for its size (normalizing the scale in grams), it becomes far more fathomable.[3]

Average Energy Rate Densities		
System	Age (Gya)	Φ_m [erg/s/g]
Human society	0	500,000
Animals, generally	.5	40,000
Plants, generally	3	900
Earth's geosphere	4	75
Sun	5	2
Milky Way	12	0.5

It's an odd relationship, isn't it? Or is it? Entropy and the second law of thermodynamics apply to closed systems. But open systems can decrease their local entropy by using incoming energy to increase and maintain the complexity of their structure and processes, even as the overall entropy of the rest of the universe increases. This is how plants grow, by drawing new energy from the sun and converting it into new structures and stored energy. In the process, they produce waste heat and noise that radiates into the ecosystem and eventually out into space.

It's a zero-sum game, played out between the conservation of energy and the second law of thermodynamics. Increasingly complex systems by their nature have lower entropy; therefore, this must be counterbalanced by an equivalent increase of entropy in the universe as a whole. At the same time, these systems must be offset and maintained by an increase in the energy they process.

The physics of this phenomenon has recently been explored by MIT physicist Jeremy England. Beginning from basic principles of thermodynamics, England shows how emergent self-organization is nature's way of dissipating and rebalancing excess energy in its efforts to attain equilibrium.[4] One way it does this is by promoting dissipative systems that can dump heat into the universe. For instance, according to England, the sun shining long enough on atoms will in time lead to their self-assembly into a configuration that best redistributes energy back into the environment as heat. The configuration goes on to organize itself with other like systems to redistribute energy even more efficiently. In this way, we have a possible mechanism by which order comes to stars, to self-replicating molecules, to single-celled organisms. These emergent, often self-replicating processes are by their nature more complex and therefore more energy-hungry than less complex ones. They can also be more inclined to proliferate over time than nonreplicating processes, thereby increasing the overall production of entropy.[5]

This is in line with an observation of Harold Morowitz, the biophysicist who studied the application of thermodynamics to living systems in the mid-twentieth century, when he said, "The flow of energy through a system acts to organize that system."[6] It seems to also mesh very nicely with Chaisson's ideas. England's hypothesis is also consistent with how, as certain aspects of the universe climb the ladder of complexity, they can become increasingly more efficient at processing energy. This plays out whether the energy is thermal, kinetic, chemical, or electromagnetic. If this really is how nature works, then at scales ranging from the atomic to the cosmic, this could be a fundamental driver of adaption and complexity.

We see this thermodynamic rebalancing regularly in far simpler systems. For instance, in fluid dynamics, a liquid that flows in parallel layers, such as water through a pipe, can be in one of two states. It is either *laminar*, in which case it is orderly and has a minimum of mixing, or, at higher velocities, it becomes *turbulent*, a state that is marked by chaotic velocity and pressure changes. This turbulent flow results in eddies and vortexes that more rapidly dissipate the kinetic energy and accelerate mixing, thereby increasing the rate of entropy production in the system by more rapidly maximizing the number of accessible microstates. This allows equilibrium to be achieved far more rapidly than it would in an orderly laminar flow.

This seems to be analogous to what occurs in other natural systems, though the specific methods of reorganization and energy transfer may differ. In the case of a star system, matter aggregates in such a way that it gives rise to emergent properties such as nuclear fusion, nucleosynthesis, and stellar evolution—the processes that create all of the elements in the universe.[7] These exceptions to the general energy flow of the universe are the product of its eddies and vortexes, metaphorically speaking, the turbulent redistribution that speeds up entropy production in its effort to balance the system.

We can also see this sort of behavior taking place on smaller, more local scales. In the case of twenty-first-century Earth, we see an increase in the number and scale of hurricanes, typhoons, and other cyclonic phenomena as global heating drives our planet's system further out of equilibrium. These turbulences are natural, energy-driven events that seek to rebalance the thermodynamics of the system as rapidly and efficiently as possible. The fact that this rebalancing can have a huge impact on human lives is of no consequence to the process.

Extending our analogy, we are faced with an intriguing implication. Turbulence in a fluid or a gas has a fractal property of generating further eddies within eddies that spread its kinetic energy and lead to equilibrium more rapidly.[8] Can we extend this model to other systems? Consider that over an extended time frame, a solar system has many scales and types of turbulent mixing taking place. These eventually contribute to planetary formation, the appearance of life, higher intelligences, technological culture, and who knows what beyond that. If each of these leads to more rapid energy processing and generation of entropy, does it eventually increase the rate at which the universe achieves equilibrium?

In "Causal Entropic Forces," Wissner-Gross explores the concept of maximizing future freedom of action, coming to the conclusion that "Adaptive behavior might emerge more generally in open thermodynamic systems as a result of physical agents acting with some or all of the systems' degrees of freedom so as to maximize the overall diversity of accessible future paths of their worlds." By exploring the maximum number of possible future paths, those paths with the greatest capacity for long-term success can potentially be realized. In this way, entropy gives rise to ever greater complexity over time.

Figuratively speaking, we are all the offspring of the eddies and vortexes of the advancing universe.

To be clear, causal entropic forcing as Wissner-Gross defines it applies to "cognitive-adaptive organisms that have an internal mental model of the environment they seek to adapt to." But if we're looking at this from a more fundamental probability-based perspective, then high-entropy states should create the best conditions for exploring the opportunity space for *any* given domain. In other words, the more unrestricted the environment, the more options that can be explored. This concept of causal entropic forcing seems like it should be critical to the origin of life, as we'll discuss in future chapters.

Given all of this, we need to ask ourselves some very big questions. Is intelligence exclusively a cognitive-based property, as we traditionally think of it, or is it much more than that? Is it perhaps even a manifestation of a much larger and universal process, one that is initially dependent on probability but, given the resultant emergent properties, becomes more and more capable of self-directed volition over time? Is every emergent progression that increases complexity and future opportunity for itself in the universe its own form of intelligence, a response to the conditions that produced it? As we progress from chemistry to prebiotic self-replication to single-cell life all the way to *Homo sapiens*, can we view each as its own emergent variation of adaptive collective intelligence relative to the precursors it evolved from? Perhaps most importantly, should we be thinking of this interpretation of intelligence as a response to and result of causal entropic principles and therefore an inherent property of the universe?

It is a lot to consider, but something well worth keeping in mind as we investigate the next stages of evolution in the universe and, most importantly for us, here on Earth.

CHAPTER 4

THE EVOLUTION OF LIFE AND BIOINTELLIGENCE

"We define ourselves as intelligent. That's odd, because we're doing the definition—we're creating our own definition and saying, 'We are intelligent!'"

—Neil deGrasse Tyson, astrophysicist, author, and science communicator

As our ship touches down on a very young Earth, it takes almost no effort to detect amino acids nearby. Virtually every stream, lake, and pool is teeming with them. Amino acids are organic compounds essential to generating and maintaining life on our planet and possibly elsewhere. Of about five hundred naturally occurring amino acids, only twenty-one are encoded by genetic material and considered essential to life as we know it.[1]

But this trip to Earth is hardly our first encounter with these organic compounds. There is growing evidence that nearly anywhere hydrogen, oxygen, nitrogen, and carbon are present, amino acids and other organic compounds can be found as well. Even meteorites and comets are frequent repositories of amino acids. For instance, on September 28, 1969, a bright fireball appeared in the Australian sky near Murchison, Victoria.[2] More than two hundred meteorite fragments were subsequently gathered and tested, revealing at least fifteen amino acids and a variety of organic compounds. In 2009, NASA announced they'd identified the amino acid glycine in a comet sample returned to Earth in 2006 by their robotic Stardust mission probe.[3] When the Rosetta space probe visited Comet 67P/Churyumov-Gerasimenko in October 2014, it found glycine as well as a number of organic precursors.[4] Perhaps most surprising, also in 2014, the complex organic compound isopropyl cyanide, which has a branch backbone essential to the formation of amino acids, was detected 27,000 light-years away near the center of our galaxy![5] Using spectral analysis, the compound was found in a cloud of gas that will one day give birth to new stars and is concentrated in sufficient quantities to be detected across that vast distance (1.56×10^{17} miles).

It appears that under the right conditions, these building blocks of life can appear almost anywhere. But what are the right conditions, and how common or rare are they?

An early insight into this question comes to us through an experiment performed in 1952 by Harold Urey and Stanley Miller, then of the University of Chicago. Nobel scientist Urey[6] and then–graduate student Miller were inspired by Alexander Oparin and J. B. S. Haldane's hypotheses that complex organic compounds may have been naturally synthesized under the primitive conditions found on Earth some four billion years ago. In what is known as the *heterotrophic theory* of the origin of life, Oparin and Haldane independently conceived[7] of a prebiotic primordial soup of basic chemicals from which life might have synthesized early in Earth's history.

Urey and Miller's experiment was elegantly simple. Sealing water, methane, ammonia, and hydrogen in a small sterile flask, they heated the mixture, causing it to evaporate into a larger flask, where it condensed and recycled, simulating the atmospheric and water cycles of our early planet. To this, they added regularly sparking electrodes to simulate lightning.

The chemicals quickly reacted with one another, and the liquid in the flasks grew pink within a day. By the week's end, the contents were deep red and brackish. Not only did follow-up testing show the development of several types of amino acids, but later, more sensitive tests revealed far more amino acids had been generated than originally thought.[8] These included not only most of the amino acids needed as building blocks for life on Earth, but also many that are not. This famous experiment (and the many reproducing it through the years) demonstrates the relative ease with which these building blocks of life are formed, a critical early step in the far longer and complex process of *abiogenesis*, the formation of life from nonliving chemicals.

It's worth looking at this again, but this time through the lens of causal entropic forcing. In a state of high entropy, this assemblage of Earth's very early chemistry was recombined into an enormous number of possible combinations, presumably under a wide range of conditions over time, until one or more held the greatest probabilistic potential for future opportunity, especially in terms of self-replication. Had this been a low-entropy system, existing levels of organization would have prevented a full range of novel opportunities from being adequately explored. By contrast, a high-entropy state allowed for a much larger range of potential arrangements and in the process found at least one with a high potential for future opportunity.

Miller and Urey's work models only one of many critical transitions in the processes that made possible what we know today as life on Earth. Each in its own way gave rise to greater differentiation and specialization in a way that would "maximize future freedom of action" within its environment. For instance, it has long been demonstrated that numerous autocatalytic sets of peptides and polymers can spontaneously give rise to self-replicating hypercycles of these molecules.[9] In chemistry, an autocatalytic set is a specific collection of elements or molecules that naturally combine into a self-sustaining, self-replicating cycle.

Theoretical biologist and complex systems researcher Stuart Kauffman has been working on autocatalytic sets since the 1960s.

Recently, Kauffman, Wim Hordijk, and Mike Steel published a paper showing how a single autocatalytic set can produce new subsets within it that are mutually interdependent on each other.[10] It is thought that such autocatalytic sets may have been the mechanism by which simple elements and compounds replicated and evolved over many millions of years, eventually leading to the emergence of life. While in and of themselves such sets are not alive, they maintain and propagate themselves in a manner that fulfills at least some of the criteria for life, suggesting this could have been an incremental and feasible path to life on Earth.

This is extremely intriguing and ordinarily would remain in the realm of speculation. But Kauffman's team showed mathematically how such sets occur naturally. Perhaps more important is their finding that this process can lead to new interdependent subsets, allowing the evolution of newer, more complex sets. According to Kauffman, this gives rise to "autocatalytic sets of autocatalytic sets, that is, true emergence."

Additionally, Kauffman maintains that the mathematics show this behavior to be substrate independent; that is, it doesn't apply only to molecules. Such autocatalytic sets might occur in colonies of bacteria or perhaps, in time, in metamorphic computer viruses. Each would potentially have the ability to form and evolve into new, emergent phenomena.

There's no knowing how long it took the early amino acids to assemble into Earth's first auto-catalyzing sets, but we do know they had a lot of time in which to do it. Many steps would have to follow, but based on our current best estimates, the first single-cell prokaryotes (early cells lacking a nucleus) may not have appeared until 3.8 to 4.1 billion years ago, around half a billion years after the Earth was formed.

There has been a lot of speculation about how these early molecules autocatalyzed their way up the ladder of complexity, going from the building blocks of amino acids to the first self-replicating life. This process, abiogenesis, needs to overcome some fairly overwhelming odds—if you're trying to duplicate it in a modern laboratory. But in the wild, across a lab the size of the Earth, processing trillions upon trillions of combinations at the same time across hundreds of millions of years, it is a very different situation. Given the right steps and conditions, it begins to feel as though life is less a slim chance and more an inevitability.

One particularly compelling model describes several stages that take us from simple chemicals to the first single-celled organism. Ammonia compounds called amines link together to form short polymer chains known as peptides or longer chains that may be polypeptides or proteins. The right combinations of these have been shown to form self-replicating loops. What this means is peptide A acts as a template to combine with peptide B and C to form a duplicate, A1. A1 is then ligated (separated from A), allowing each chain to proceed as a template for further replication. As this cycle builds and grows in complexity, a cyclic, autocatalytic set of processes ensues, which can potentially develop into what is known as a *hypercycle*. This process undoubtedly would have to occur countless times and across many stages of increased complexity before giving rise to early versions of RNA, ribonucleic acid, a molecule found in all living cells on Earth. At some point, among the many variants of RNA, a ribozyme[11] formed that could act as both a gene encoder and an enzyme (catalyst) to drive and define protein synthesis. From the point when RNA arises, given enough time, organic molecules might have gone on to self-assemble abiotically into droplets forming a protobiont, a collection of autocatalytic organic molecules contained within a membrane. A protobiont would have been a precursor to a true prokaryote, a simple self-replicating cell such as a bacterium, one of the first forms of life.

A major obstacle for this hypothesis is how cell membranes came to exist. Such structures are considered essential to the development of early single-celled life, and it is challenging to see how one would materialize spontaneously in the primordial soup of early Earth. Modern-day cell membranes are dependent on the unique bonds in amphipathic molecules[12] known as phospholipids. Depending on their orientation, phospholipids can be either hydrophobic (repelling water molecules) or hydrophilic (bonding to water molecules). It's been conjectured that if enough phospholipids accumulated, they could self-assemble into a lipid bilayer, a double layer of fatty acids that renders the membrane hydrophobic in some regions and hydrophilic in others.[13] In other words, this makes different parts of the cell's membrane permeable to water soluble molecules.

One hypothesis of how and where these molecules would aggregate to form such a membrane comes from German chemist Günter Wächtershäuser. His concept of a *primordial sandwich* (Primordial soup. Primordial sandwich. Who says scientists don't have a sense of humor?) proposes that a sufficient concentration of amphipathic molecules could have accumulated on the surfaces of minerals. There, they would have self-organized into a simple bilayer that could have been a precursor to true cell walls.

There is at least one problem with this hypothesis, which is that while phosphorus is essential to biology on Earth and abundant in the universe, it's far less common on this planet. Moreover, a great deal of our world's phosphorus is locked up in the form of minerals that are inaccessible to living things. So where did it come from?

Many scientists have speculated that bioavailable phosphorus came to us from outer space, but only recently has this been substantiated. In 2018, a team of researchers from the University of Hawaii at Manoa, in collaboration with colleagues from France and Taiwan, performed ultra–high vacuum chamber experiments that replicated conditions we might expect on icy grains of comet and meteorite dust coated with carbon dioxide, water, and phosphorus. Exposed to the equivalent of the high-energy cosmic rays found in space, reactions occurred that resulted in "biorelevant molecules such as oxoacids of phosphorus."[14] Such a reaction would have been essential to the beginnings and continuation of life because phosphorus is an essential building block in biology. It is especially important for energy transfer in cells in the form of adenosine triphosphate (ATP), as well as for the lipids found in cell walls. So, we can be thankful for all of those meteorites that slammed into the Earth as it was forming 4.54 billion years ago.

The reason cell walls are so important is that they allow a sufficient concentration of organic molecules to accumulate within them for cyclical chemical reactions to occur. As a result, contained autocatalytic cycles can be established. Over long periods of time, the tiny lipid-shelled containers, or cell sandwiches, essentially become billions and trillions of mini-laboratories performing a vast array of experiments. Even though they weren't alive, those proto-life-forms—or protobionts—that were most successful could go on to proliferate. It's not true Darwinian evolution, but the beginnings of something similar.

The formation of life on Earth took place some four billion years ago during the Archean Eon, which came immediately after the Hadean Eon that followed the formation of our planet. The Archaean Eon is deemed the earliest period in which conditions on Earth would have been sufficiently hospitable to allow life to develop and thrive. Based on present-day evidence, these conditions must have eventually given rise to autocycles of nucleic acid, self-replicating RNA, and in time, deoxyribonucleic acid (DNA).

Obviously, the exact sequence of events that allowed simple elements to become self-replicating life can only be surmised, but the diagram on the next page paints a fairly reasonable picture that gets us over many conceptual humps.

The diagram depicts a simple progression, but life is not a straight line and the beginnings of life were no different. Odds are that dozens, perhaps hundreds or thousands of autocatalytic strategies arose that could have potentially given rise to life in time. But just as it has been observed that history is written by the winners, so too, our knowledge of early pre-life is mostly limited to what we can learn from and hypothesize about the survivors. Nothing remains of the earliest autocatalyzing chemical reactions that did not go on to succeed and survive into the present day. Nevertheless, we may one day manage to run enough computer simulations on those primordial chemicals and conditions to discover a list of likely candidates.

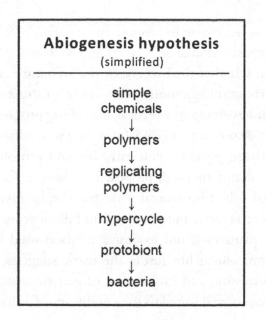

Abiogenesis hypothesis
(simplified)

simple
chemicals
↓
polymers
↓
replicating
polymers
↓
hypercycle
↓
protobiont
↓
bacteria

Perhaps the earliest of the microbial lines that survived to the present day was recognized only very recently. In 1977, American microbiologist and biophysicist Carl Woese defined the domain Archaea based on his study of the evolutionary history of the earliest prokaryotic organisms.[15] Previously, it had been thought that life originated in two domains, the prokaryotes, generally bacteria, and the eukaryotes, such as protozoans and algae, and much later the animal kingdom, including ourselves. The primary distinguishing trait between the two domains is that prokaryotes do not have a nucleus for containing and protecting their DNA or other organelles, among them mitochondria. These microcompartments separate various regions and processes in the cell from other parts, facilitating a degree of specialization within each part. But then Woese discovered that some of the prokaryotic microbes he was studying were far more similar to eukaryotes (and therefore ourselves) than they were to traditional prokaryotes. Despite lacking a nucleus like prokaryotes, these microorganisms have genes and a number of metabolic pathways, including transcription and translation (the means by which genes communicate their instructions for protein production), that make them much more related to eukaryotes than to

prokaryotes. This led Woese to propose the separation of the prokaryotes into two separate domains: Bacteria and Archaea.

It's uncertain whether the bacteria or the archaeans came first, but
it may be an irrelevant question. We've come to think of organisms
perpetuating and evolving in a vertical, branching progression, such as
we see in the transmission of DNA from parent to offspring, but this
wouldn't have been possible with early life and protolife. However,
there's another major method, proposed by Woese, for exchanging
genetic material called horizontal gene transfer, or HGT.[16] HGT is
still used by bacteria today, but three or four billion years ago, it would
have been the primary if not exclusive method used by all protolife and, later, unicellular life. Just as the name suggests, this involves
the lateral transmission and exchanging of genetic material between
organisms, a process well suited to the conditions of a shared "primordial soup."

However, cellular evolution through HGT would have been
self-limiting, because as an organization becomes increasingly complex, the ways it can successfully be changed through such recombination diminish. According to Woese, "Increased integration and
complexity buy specificity, but at the cost of flexibility." He goes on
to say, "Primitive cellular evolution is basically communal. . . . It is the
community as a whole, the ecosystem, which evolves." Woese maintained HGT was the primary method of gene transmission until some
cell lines reached a critical level of complexity and transitioned from
being "completely ephemeral to being increasingly permanent." At
this point, life crossed a line Woese called the "Darwinian threshold,"
when speciation as we think of it came to exist.

This critical moment in the development of single-celled organisms marks the beginning of the universal tree of life, the first juncture
at which we can represent the evolution of organisms in such a manner. If Woese is correct, and I think he is, the tree of life has no root,
no truly knowable point of origin, other than the Darwinian threshold. By extension, it's unlikely there was ever a true Last Universal
Common Ancestor (LUCA) for the three domains, Bacteria, Archaea,

Phylogebetic Tree of Life

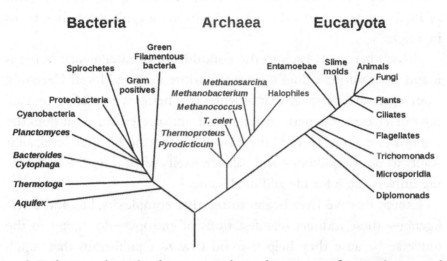

Bacteria   **Archaea**   **Eucaryota**

and Eukarya, though there may have been one for Archaea and Eukarya.

Once again, we find ourselves on a new threshold as an aggregation of basic elements combine in conditions of high entropy to explore an enormous range of possibilities that results in a new form of emergent complexity. The difference this time is that this process has actually resulted in life, and with it we are a step closer to what we traditionally think of as intelligence.

But was this advance truly that different from the prior emergent responses to entropy that we have witnessed? If conditions give rise to a new order of complexity that is better able to proliferate, while at the same time reducing its local entropy, could that be considered a form of intelligence relative to what came before?

Perhaps it's time we looked at this from a rather different perspective.

————

Our ship leaps forward into the present day to find a world teeming with people—seven and a half billion of us. This globe-spanning species of ours is arguably one of the most prolific and consequential this planet has ever seen. Only bacteria, which account for more biomass than all of the plants and animals combined, might be considered as successful, if not more so. We, of course, have the upper hand in terms of intelligence, but that doesn't mean bacteria can't regularly take us in a fight.

Which brings us back to the conundrum of intelligence. What is it and what purpose does it serve at a more universal level? Certainly, from a personal perspective it provides us a better means of perpetuating ourselves, our genetic material, and our species, but what does the cosmos care about that? At the most basic thermodynamic level, all it wants to do is equilibrate and balance itself. What does the unthinking universe care for life and intelligence?

Except that we have begun to see that complexity, life, and intelligence—these indirect manifestations of entropy—do matter to the universe because they help it trend toward equilibrium that much more efficiently. As life gave rise to more complex and intelligent species, it set the stage for a new form of complexity and intelligence: human society. The swiftly developing aggregation of organisms that is our global society is rapidly climbing the wall of complexity, even as it generates heat far faster than our planet is able to radiate it. If history is any indicator, and it almost certainly is, this progression may soon lead to a new emergence, one that is orders of magnitude more complex and efficient at processing energy.[17]

But we'll explore that eventuality in the coming chapters. For now, let's return to the important question suggested above: where and when did intelligence begin? To consider it, we'll need to retrace our steps.

It's pretty well established that our mammalian ancestors exhibited differing degrees of intelligence, which is generally attributed to their having a neocortex, the most recently evolved part of the brain's cerebral cortex. Made up of six layers of specialized neurons, the neocortex appears to be unique to mammals and contributes to many higher-order brain functions we closely associate with intelligence. Even some of the earliest mammals, such as the lemur-like strepsirrhines, have been shown to be capable of understanding numbers, abstract concepts, and risk. These early prosimians evolved sixty-three million years ago, so presumably the headwaters of intelligence go back much further.

If we step back 100 million years before the strepsirrhines, we find ourselves in the age of dinosaurs. While we can only infer the behavior of these reptiles from the fossil record, their relatives that have survived into the present day may provide some indication as to their intelligence. Lizards known as anoles are one well-studied species whose lineage directly connects them to the era of the Jurassic giants. Despite their diminutive size, research studies have found anoles to be far more intelligent than we might expect from a reptile. Tests performed at Duke University have shown them capable of retrieving an insect hidden beneath an opaque colored disc.[18] Changing the number and colors of lids showed not only that the anoles were capable of learning, remembering, and adapting, but that they also learned more quickly than birds tested in similar studies.

Jumping back another eighty million years delivers us to a time before the dinosaurs, when crocodylomorphs walked the earth. While few of their crocodile descendants have been the subjects of extensive intelligence testing due to their size and ferocity, the smaller Cuban crocodile has been the subject of several studies. These creatures have been found to use a range of signaling methods including posture, vocal sounds, infrasound, and head slaps to communicate with one another. Several types of play have also been observed, another hallmark of intelligence. Though we typically think of this species as being a solitary hunter, cooperative behavior in the form of pack hunting has occasionally been witnessed. Learning and tactics can probably be added to this list, since there have been numerous sightings of crocodiles balancing sticks and twigs on their snouts to attract nest-building birds for an impromptu lunch date. All from a reptile with a brain the size of a walnut weighing less than half an ounce!

Evidently, we'll need to consider still earlier animal lineages, but first let's take a look around the Triassic swamp forest that is the crocodylomorphs' habitat. While it's still millions of years too early to find any flowering plants, that doesn't mean there isn't plenty to see and consider.

Conifers, ferns, horsetails, and other spermatophytes, or seed plants, dominate the land. Can we consider these plants to be intelligent too?

Current-day popular culture has embraced the idea of plant intelligence for several decades, though much of this belief has been based on less-than-rigorous scientific research. In contrast, evolutionary biologist Monica Gagliano, who has adopted methods from animal behavior science, maintains that plants also possess intelligence and memory as well as the ability to learn and to communicate with neighboring plants.[19] While Gagliano's work has attracted some criticism, much of which is around terminology, her research has been solid enough to receive peer-reviewed publication. Gagliano acknowledges that plants don't have neurons and that learning and intelligence in plants isn't the same as for animals. But why should we think all life should be intelligent in the same way? If something is able to adapt to its environment, signal or exchange information with its kind, and learn from experience, all as a means of promoting future freedom of action and thereby perpetuating its survival and that of its species, how is that not intelligence simply because it doesn't occur through the same electrophysiological signaling we see in the animal kingdom?

Continuing our retro-journey, we jump back to 530 million BCE,[20] the midst of the Cambrian Explosion, a period that saw a tremendous increase in the diversity of new species and body plans. Lasting a mere fifteen to twenty-five million years, this was the era when most of the major animal phyla first emerged.

One predominant species during this era were the ammonites, amazingly constructed creatures with hard shells formed into logarithmic spirals. Their shells contained multiple chambers that protected their soft bodies and allowed these creatures to control buoyancy as they jetted through the oceans. Though very similar in body plan to the modern-day chambered nautilus, ammonites were actually more closely related to another branch of the cephalopods, the squid and octopus. We don't have a direct way to gauge the intelligence of these ancient creatures, but we do know a great deal about their descendants. Considered to be among the most intelligent of all the invertebrates, squids and octopuses have been shown to engage in social communication, tool use, learning, and problem-solving, all hallmarks of intelligence. This shouldn't be surprising. The ammonites didn't survive for nearly 200 million years without the ability to adapt to their environment.

Traveling back another fifty million years delivers us to the time of some of the earliest multicellular organisms: the Cnidaria, a phylum that includes the hydra. Measuring less than half an inch in most species, hydra look like nothing so much as an elongated sack ending in a mouth opening surrounded by a number of thin tentacles. Yet despite its diminutive size, the hydra remains a fearsome predator that feeds on rotifers, insect larvae, small crustaceans, and annelid worms. While the hydra is fascinating for many reasons, perhaps the most important feature from our standpoint is that its nerve net is considered the precursor to our own central nervous system. Though hydra lack a brain or any form of cephalization or head region, this nerve net allows it to sense its environment through physical contact so that it can detect and capture food.

While our evidence for intelligence may appear to be growing more tenuous, I would submit that there isn't a well-defined delineator so much as a spectrum—a long and broad spectrum that encompasses a very large number of responses to how best to maximize future freedom of action.

With one last jump, we arrive in a world unlike any we've visited so far. The air is anything but breathable, since it's composed mostly of methane, carbon dioxide, hydrogen sulfide, and ammonia, much like the conditions in Miller and Urey's experiment. This is the era of cyanobacteria, nearly three billion years ago. Among the earliest unicellular organisms, cyanobacteria are the dominant and perhaps most complex organisms on the planet at this time.

Cyanobacteria remained the apex species on our planet for perhaps more than a billion years, even though the Earth changed dramatically throughout that time.[21] Converting sunlight and carbon dioxide into energy and waste calcium carbonate and oxygen through the recently evolved process of photosynthesis, these organisms gradually sequestered much of the carbon dioxide in our atmosphere, replacing it with the oxygen we know and love today. Once oxygen in the oceans and atmosphere had risen to sufficient levels, perhaps somewhere between 3 and 10 percent of our modern concentrations, it became possible for higher-energy organisms to appear. This is because oxygen allows much higher-energy chemical reactions to occur, which in turn make it possible for organisms with higher metabolisms to evolve. With these changes, we begin to see predators like the hydra appear. The Cambrian Explosion was relatively soon after that.

You may think that a single cell can't be intelligent, but consider that unicellular organisms do show volition and adapt to their environments, even if not through cognitive means. Perhaps more importantly, these organisms are anything but simple. Individually, many single-celled organisms are capable of changing their physiology and behavior by altering the regulation of their gene expression so as to adapt their production of proteins or enzymes and better conform to the new conditions. Many of these organisms frequently respond to nearby chemicals, moving in the direction of the most promising environmental conditions and away from unfavorable ones. This process, chemotaxis, allows microbes to sense changes in chemical gradients, aiding in their foraging for food.

Amazing as all of these cellular processes are, the most interesting and consequential cell behaviors occur when these organisms act collectively. We tend to think of microbes individually, as if they were operating as a much-later-evolved member of the multicellular animal kingdom. In actuality, single-celled organisms thrive best in a more communal form, as members of colonies, biofilms, and slime molds. Recent studies of the slime mold *Physarum polycephalum* have even demonstrated their ability to transfer and retain acquired memory from cell to cell.[22]

Extending the idea of community still further, the soil bacterium *Bacillus subtilis* is able to form what's known as a biofilm, an aggregation of cells held together by an extracellular matrix that allows them to move and feed together. In its search for nutrients, the biofilm will expand and move toward richer nutrient sources and away from poorer ones, guided by the molecules in the environment or by the pheromones it secretes as it goes along. As this expansive growth takes place, individual members in different regions of the film will differentiate. Some cells along the leading edge may develop extra flagella,[23] which allow the colony to move more rapidly, while other members may secrete chemicals that help the colony to slide more efficiently. When especially nutrient-rich environments are found, the biofilm may even swarm outward in a surge of branching columns that get it to the source faster. All of these methods evolved so that the colony could outcompete other microbial communities.

Cells engage in communal behavior to promote the continuation of the colony by using chemical signaling, a form of cell-to-cell communication and cooperation. Under certain conditions, cells can act mutually for the benefit of each other or even altruistically,[24] promoting the survivability of other cells instead of themselves.

The various emergent behaviors we've discussed—abiogenesis, unicellular gene expression, communal microbial intelligence, and multicellular differentiation—seem very different on one level, yet they can also be seen as manifestations of the same pattern: collective, emergent intelligence driven by competition, thermodynamics, and entropy as a means of promoting future freedom of action. Viewed this way, the fundamental processes of the universe seem to keep producing similar outcomes again and again. Will the trend continue for the foreseeable future? Perhaps a better question to ask would be, after 13.779 billion years, why should we expect this pattern to suddenly change now?

As to the underlying process behind this pattern, it's time to take a deeper look at what is driving our universe toward ever greater levels of complexity and intelligence.

CHAPTER 5

COMPLEXITY AND EMERGENCE

"We are agents who alter the unfolding of the universe."
—Stuart Kauffman, MD, biologist
and complex systems researcher

The heating and cooling of the Earth's oceans and land masses gives rise to sustained vortexes of air and moisture, resulting in a globally interconnected weather system. A single ant exhibits a series of relatively simple behaviors, yet when those same behaviors are performed by thousands of similar ants, far more complex activity results, allowing them to forage for food more effectively, build bridges constructed from nothing but their living bodies, and create elaborate fractal-like subterranean nests. A solitary neuron is capable of several extraordinary functions that are relatively meaningless in isolation, but put a sufficient number together in the right combination and place them in a sensorially enriching environment, and they can self-organize and interconnect according to rules we are only still discovering, leading to abstract thought, self-awareness, and even consciousness.

What do the weather, ant colonies, and brains have to do with each other? They are all manifestations of complexity and emergence, terms that are still being defined by researchers in the field of complexity science.[1] But as challenging as understanding these phenomena can be, emergence is basically what happens when the behavior of a system is greater than the sum of its parts.

Sound familiar? This is essentially what we've seen in so many of the systems we've visited since the Big Bang. Atoms that fuel stellar evolution and nucleosynthesis. Elements that accumulate to form planets with geologically, sometimes even ecologically active systems. Simple molecules that give rise to self-replicating, autocatalytic polymers and hypercycles. Protobionts that lead to unicellular organisms. Colonies of single-cell organisms that eventually evolve into multicellular organisms. Only in retrospect and with considerable knowledge and tools do these transformations seem even moderately similar or comprehensible.

But the reality is that this is the very nature of emergent behavior: to give rise to something completely unexpected from the collective behavior of far simpler constituent parts and actions.

How did complexity science and the study of emergent behavior come about? Historically, much of scientific inquiry has relied on a reductionist approach, the idea that systems can be studied and understood entirely in terms of their simpler, more basic components and principles. This idea goes back at least as far as the Greek philosopher Aristotle (384–322 BC). However, the approach was firmly established with the onset of the Enlightenment in the late 1600s. The philosophers and scientists of the Enlightenment believed that every aspect of the natural world was deterministic and could be reduced to its constituent parts. (One exception was the dualist belief that the mind and consciousness somehow originate via other natural or supernatural means.) While this approach worked well enough for the early stages of formalized scientific inquiry, by the late nineteenth and early twentieth centuries it was becoming apparent that it had significant shortcomings. Einstein's theories of special (1905) and general relativity (1907–1915), the advent of quantum mechanics beginning around the turn of the twentieth century, and Gödel's incompleteness theorems (1931) set the stage for a new era in which the limits of reductionism and determinism were increasingly recognized.

The irreducibility of so many complex systems, from the weather and climate to economies and the stock market, was a tremendous challenge to the reductionist approach. The challenges of understanding such complex systems plagued researchers across many different fields for decades. Then in 1984, two dozen notable scientists and researchers, including Nobel laureate Murray Gell-Mann, gathered in Santa Fe, New Mexico, for an interdisciplinary workshop on "Emerging Syntheses in Science." From that early seed, the now world-famous Santa Fe Institute was born.

Dedicated to "understanding and unifying the underlying, shared patterns in complex physical, biological, social, cultural, technological and even possible astrobiological worlds," the Santa Fe Institute (or SFI) is an independent theoretical research institute that stands at the forefront of complexity science and the effort to understand complex adaptive systems and emergent processes. While there had been other institutes that researched these phenomena, such as the Institute for Advanced Study at Princeton, at the time, SFI was unique in its radical interdisciplinary approach.

Stuart Kauffman, an early faculty member at SFI, has spent the better part of his career exploring the nature of complexity and emergence. According to Kauffman, true "radical emergence" is unique in that "we can't pre-state it, we can't know how it happens, and it changes the course of evolution." This implies that many if not all natural processes[2] that are emergent somehow contribute to evolution in the universe, be they stellar, planetary, autocatalytic polymers, interdependent collectives of single-celled organisms, or technological societies.

However, it is important to remember that complexity in and of itself is no guarantee or predictor of emergence. There are plenty of systems in the world that can be said to be complex, yet this is not enough in itself to lead to emergent behavior. True emergence arises because the *collective interactions* between the constituent parts of a system result in behaviors that are unforeseeable based on the function of those parts.

What is it about aggregations of complex elements or "agents" that leads to increasing complexity and emergence? How and why does it happen, and do all agglomerations behave this way? The answer to this last item is a resounding "No." Obviously, if you assemble thousands of pencils or paper clips, they will not spontaneously increase in complexity and lead to some new form of emergent behaviors. This may be because one of the keys to emergence is not in the quantities or even the connections (though these are important) but in the communication, or, more accurately, the structured patterns of cooperative interaction. Atoms that exchange electrons and forces according to universal, immutable rules as they seek their minimum energy states. Cells that communicate and cooperate with each other using chemical messengers, including hormones and neurotransmitters, as well as quorum sensing (a type of regulatory process that allows cells to cooperate based on the density of the sensing population). Termites that share pheromones that direct their activity. As individuals, each of these agents follows a very simple set of rules, but in much larger numbers their rudimentary behaviors give way to unexpected phenomena. Atoms lead to the highly structured table of elements and molecules that are virtually limitless in form and properties. Termites generate solar-cooled cathedral-like mounds reaching more than ten feet high.[3] Relatively simple cells beget everything from hydra to human beings. All from a relatively small number of fundamental rules that are continuously transmitted between each member.

Swarming behavior is an excellent example of this. If you've ever seen a murmuration of starlings, you have been witness to emergent behavior. A murmuration is a flock of thousands, sometimes even tens of thousands of birds that manifests as an organic mass that seems to move and flow as though it is a singular entity. It is beautiful to experience, but all the more so when we realize this phenomenon is generated by each bird adhering to a very basic set of rules. In 2013, researchers from Princeton, the University of Toronto, and Sapienza University of Rome published the results of their studies of still images taken from videos of starling murmurations.[4] Creating a mathematical model and using systems theory, they showed that starling swarms are most cohesive and energy efficient when each member attends to seven of its neighbors. By maintaining approximately the same distance and speed relative to one another regardless of which way they move, these birds continuously realign their flight in coordination with each other, thereby precipitating an emergent event.

This observation was modeled years earlier in 1986 by *artificial life* researcher and computer graphics expert Craig Reynolds.[5] Artificial life is the study of the dynamics of living systems and populations through computer simulations, and one of the earliest of these was Reynolds's flocking simulation called "Boids." In the simplest version of this program, individual boids followed three simple rules: (1) try to maintain a minimum distance from other objects in the environment, including other boids; (2) try to match velocities with immediate neighbors; (3) try to move toward the perceived center of mass of its neighbors. The result is a graphic visualization of swarming behavior that emulates a flock of living birds with stunning authenticity. Such swarming is a model of *collective intelligence*, in which there is no hierarchy or form of top-down leadership. There is only the behavior of thousands of individual agents interacting as equals, all adhering to the same limited set of rules that result in something utterly unique and beautiful.

How universal are such phenomena in the general sense? The field of artificial life, or A-Life, seeks to show that such emergent behavior needn't be limited only to living things, or even to natural ones for that matter. In the 1940s, physicist and computer pioneer John von Neumann sought to model self-reproducing machines using cellular automata (CA), the invention of his close friend and colleague, Stanislaw Ulam. A CA is a matrix of "cells" that follow a particular rule or set of steps. While von Neumann managed to design such a rule with considerable effort, it was complicated, requiring twenty-nine steps during each generation.

Then, in 1970, Cambridge mathematician John Conway invented "The Game of Life," in which a cell is said to either be alive (on) or dead (off).[6] "Life," as it is commonly known, follows a simple set of rules on an infinite grid of cells.[7] Which state each cell transforms into is determined by the current states of its neighbors, those cells that are horizontally, vertically, and diagonally adjacent to it. For some rules, with each passing generation, cells appear to move and flow across the matrix, sometimes in repeating patterns, sometimes as massive fractal objects. Other rules quickly stabilize or rapidly go extinct.

Fascinated with the implications of CA, in the 1980s, computer scientist and physicist Stephen Wolfram set out to catalogue and study all possible combinations of the simplest form of CA, known as elementary cellular automata.[8] This special class of CA is considered one-dimensional because it takes place on a single line of eight cells with each cell's state determined by the state of the two neighboring cells on either side of it. There are 8 (2^3) possible configurations for each cell and its neighbors, resulting in 256 (2^{2^3}) possible elementary cellular automata. Since several CAs yield patterns that are equivalent or mirror images of each other, there are effectively eighty-eight elementary CAs that perform unique transformations, or Rules. As the cells are transformed with each passing cycle, a pattern emerges that Wolfram classified into four different classes: (1) those Rules that quickly settle into a uniform, unchanging pattern, including those that go extinct; (2) those that result in a uniform pattern or cycle between patterns, depending on their starting state; (3) those that produce a seemingly random state with the appearance of occasional structures; and (4) those that are a mixture of order and randomness with structures that move about and interact in complex ways, over long periods of time.

Numbering the Rules according to his own binary-based system, Wolfram was especially fascinated by Rule 110, which exhibits interesting Class 4 behavior on the boundary between order and chaos. Despite being generated from the simplest of instructions, Rule 110 results in structures that appear and interact in intricate ways. Wolfram conjectured that this rule was universal or Turing complete. Such a universal Turing machine is capable of manipulating data to perform any type of algorithm (instructions) much as modern computers do.[9] In 1994, Wolfram and his assistant Matthew Cook proved that Rule 110 was in fact Turing complete, making it the simplest universal Turing machine ever created.[10]

What does all of this have to do with our search for intelligence in nature? Well, Wolfram speculates that most if not all Class 4-type behavior is Turing complete, meaning this could be very common in all aspects of nature, not just computer automata. Though Wolfram's thesis has been considered controversial among many scientists, it points to some highly intriguing possibilities in the natural world based on the interactions of very simple relationships. It seems to me unlikely that nature, life, and thought are computational in the true digital-binary sense of the word, but even if they aren't, the relative ease with which emergence appears based on only a few simple rules should give us pause. If universal computation can be achieved based on such a simple system, what does that mean for naturally occurring and evolving complexity and intelligence? Does increasing complexity hide or mask this computational nature from us? Could life be considered a form of computation, even if it isn't necessarily digital or Turing complete? And if so, could a sufficiently complex computer one day achieve or even surpass humanity in its ability to grasp, analyze, reflect on, and introspect about the universe around us? The nature of such a potential emergence is as fascinating as it is existentially concerning and will be explored throughout the rest of this book.

Then there's the matter of this mysterious boundary between order and chaos, exhibited by Rule 110 and other Class 4 CAs. Is there something that draws these CAs to this abstract region, and why are emergent properties so likely to appear when systems perch right on that cusp? As it happens, this boundary between order and chaos may be far more universal than one might expect. In his 1992 book, *Complexity: The Emerging Science at the Edge of Order and Chaos*, M. Mitchell Waldrop writes about the insights of computer scientist Christopher Langton, one of the founders of the field of artificial life.

> In between the two extremes [order and chaos] . . . at a kind of abstract phase transition called the edge of chaos, you also find complexity: a class of behaviors in which the components of the system never quite lock into place, yet never dissolve into turbulence, either. These are the systems that are both stable enough to store information, and yet evanescent enough to transmit it. These are systems that can be organized to perform complex computations, to react to the world, to be spontaneous, adaptive, and alive.

More and more in recent years, complex systems researchers have focused their attention on this zone they call the edge of chaos, repeatedly confirming its uniqueness as a breeding ground for emergent activity. This edge isn't a physical location but an abstract region that exists at the boundary between order and chaos. According to many researchers, this is where complexity is born.

Though several researchers beat around the edges of this concept during the 1960s and '70s, it was Christopher Langton, one of the founders of the field of artificial life, who took this idea to the next level in the late '80s. Studying Wolfram's four CA classes, Langton eventually discovered something astonishing. As mentioned earlier, Class 1 CAs are highly stable, with many of them dying out completely in a single step because every rule results in death for each cell. Langton gave these Class 1 CAs a value of 0.0. At the other extreme, Langton assigned the most chaotic Class 3 CAs a value of 0.50. This parameter, which he dubbed *lambda*, was calculated according to the probability a cell would be alive in the next generation. When Langton looked at all of the Class 4 CAs, those exhibiting the most interesting and complex behaviors, he found their lambdas were extremely similar—all hovering near a value of 0.273.[11] It seemed that the ideal systems were those that were stable enough to maintain their integrity but fluid enough to be triggered or altered in critical ways by the smallest of changes. In many respects, this describes how many naturally emergent systems operate, from murmurations to ant colonies to cognition. Poised on a knife's edge, these systems can rapidly shift and flow according to their changing environment, based on integrated, bottom-up control.

This isn't to say there's something especially significant about Langton's lambda value. It's merely a calculation that identifies a relatively consistent threshold of complexity between order and chaos for this particular system. But it does demonstrate how this edge of chaos can be identified, and sure enough, Langton and many other researchers have done this for numerous other systems over recent decades.

According to many complexity scientists, these systems give rise to a self-organized criticality that tends to drive them to the greatest sustainable level of complexity. Then, as they approach this point, either physical conditions or a competitor in the environment imposes a limiting condition and impedes further progress. A classic example of self-organized criticality is a pile of sand that is added to grain by grain.[12] The grains self-organize, interlocking in many different ways, but at some unforeseeable point, one grain too many is added and the relationship between gravity, weight, and friction make a sharp transition—in physics, what's known as a phase transition—and it results in a landslide. This landslide may be large or small according to power laws that correlate how often it happens with the size of the collapse. Small slides occur many times more frequently than larger ones. As the grains continue to be added, they accumulate and rebuild the pile until another slide is triggered, and so on.

In physics, a criticality or critical point is when a system reaches a phase boundary at which a phase transition occurs. A classic example is water transitioning between liquid and solid states at 0° C.[13] For dynamical systems, self-organized criticality is the tendency to move toward a state of maximum allowable complexity, beyond which point a phase transition occurs. Some complex systems researchers speculate that in evolution this drives genetic adaptation toward the edge of chaos, where many of the most complex and interesting behaviors can occur, including life. Since the concept of self-organized criticality was first proposed by a team led by Danish theoretical physicist Per Bak,[14] many systems that seem to exhibit this have been identified, including sand piles, earthquakes, forest fires, rivers, mountains, cities, literary texts, electric breakdowns, motion of magnetic flux lines in superconductors, water droplets on surfaces, and human brains.[15]

Such processes can also be seen in the coevolution of predators and prey or competitors vying for a niche or resources. Generally speaking, an organism is optimally balanced for the ecological niche it occupies, until a better-suited competing species or predator comes along. Because its traits hover near a criticality, over long spans of time, the species is able to evolve a trait that restores its competitive advantage. However, its upper hand disappears once its adversary evolves a countermeasure or a better version, and so on.

For instance, a particular species of frog may feed on an insect that subsequently evolves an unpleasant secretion that makes it unpalatable to the frog, promoting its survival. Then, over time, that frog species (and only that species) may evolve the ability to tolerate the taste of the offending insect, increasing its access to food and thereby improving its survivability. This process can continue to escalate, turning it into nature's version of an arms race. Each species is driven in the direction of greater complexity, but this is only possible if genetics, epigenetics, and other processes are not overly stable, but able to shift and adapt over time as they are presented with new information about their environment. A classic example of this mechanism is sexual recombination, which made the evolution of more complex organisms possible through the reshuffling of genetic material. In many respects, the drive toward complexity in the face of entropy might be seen as a more general principle underlying all biological evolution.

Another example of this fluid, edge-of-chaos behavior is cognition.[16] As a simplified example, a group of neurons in a child's brain acquire the ability to recognize the fascinating pattern that represents a bee. One day, upon touching a bee, the result is a painful sting that is registered by another cluster of neurons, which quickly become associated with the bee-image cluster. The next time a bee is seen, that associated memory is attached to it, resulting in avoiding a repeat of the painful experience. Such learning is possible only if that mind is neither too rigid nor too chaotic, but in a state in which each neuron hovers on the cusp of criticality, allowing it to quickly respond to new information. And so it seems for many other manifestations of complexity.

But there is another recurring pattern that consistently appears: the rate at which these different levels of emergence and complexity record and process *information*. The ability to act on and learn from information in the environment is critical to the success of every intelligence. According to David Krakauer, president of the Santa Fe Institute, this can be seen throughout the history of evolution. The evolution of epigenetics, switching individual genes off and on in response to the environment, was driven at least in part by the limits at which sexual reproduction could respond to changing conditions. Nervous systems, and later cognition, developed to deal with changes that occurred even more rapidly than the epigenome could respond to and learn from them. Similarly, cultural learning and knowledge transmission were a means of exceeding the limits of individual minds.

All of these forms of information processing—genes, epigenetics, nervous systems, brains, and culture—function more rapidly than the methods that preceded them as a response to competition in the environment. Each innovation adapts and learns at a different rate according to its function. The ability to respond more rapidly ties in neatly to the concept of maximizing future freedom of action. Interacting with and operating on information is the most basic and universal function of life. Improving on this capacity and speeding it up is the surest way to outcompete rivals for your niche in order to survive. Each successive stage of increased processing rates is driven by the physics of information. As Krakauer points out, if the rate at which a system records information isn't at least as rapid as the rate at which that information is changing, some information will eventually be lost. For intelligences able to benefit from improvements in *temporal resolving power*, this would be a very important driver.

Let us consider what all of this means for our general definition of intelligence. Emergence and intelligence, as broadly defined here, appear to be products of increased complexity and criticality. However, not all emergences maximize future freedom of action and result in new intelligence. So, since we may define intelligence to include living systems of differing levels of complexity (e.g., the cells in a multicellular organism as well as the organism itself), what of the underlying emergent processes that generate the intelligence, even those processes that are not themselves biologically based? Given this relationship, we may one day decide that all radical emergences that operate on and process information can be said to exhibit intelligence relative to their precursors, and that each level of these emergences likely follows a set of power laws related to the rate of energy processed, response time, and degrees of complexity. This suggests a new variant on our definition of intelligence: *Any radical emergence that decreases entropy (locally) and increases its own future freedom of action relative to those constituents that preceded it—whether they be chemical, biological, or cognitive—can be said to be a type of intelligence.* These radical emergences set the stage for generating new niches that facilitate the further development of complexity, emergence, and intelligence, the entire process driven by the universe's relentless trend toward equilibrium.

By this reasoning, intelligence isn't just a byproduct of certain emergences; rather, what radical emergence manifests *is* intelligence, at least according to our broader definition. Many emergent behaviors and systems amount to a form of collective intelligence relative to that system's constituent parts. That we human beings have the ability to ponder and reflect on these processes is simply the good fortune of where we find ourselves on the spectrum of radical emergence and collective intelligence, so we probably shouldn't be too smug about it. To future emergent intelligences, our abilities are likely to seem elementary, perhaps even quaint.

Given this, we might anticipate a number of characteristics about the emergences and stages of future intelligence that will succeed us, including their complexity, learning rate, and energy consumption. As each new emergence appears, it will typically explore as broad a domain as conditions allow, occupying and contributing to whatever ecosystem it may inhabit. We have already seen this occur in chemistry, unicellular life, multicellular life, neural-cognitive organisms, and our own socioeconomic technological world. Each of these emergences expands to occupy the many possible niches of its domain, incrementally setting the stage for future ones. Moreover, each increases the new ecosystem's number of accessible microstates, enabling the eventual promotion and generation of subsequent new emergent phenomena relative to the prior levels of complexity.

The pattern of emergence repeats again and again, and there is no reason to think it won't continue far beyond the emergences of human beings and our civilizations. Based on the past, we should probably also anticipate other trends: the continued exponential increase of energy rate density in emergent systems of greater and greater complexity, increasingly rapid learning and response times, and the reduction of the time frames in which these new levels of emergence may appear. The amount of energy harnessed and used by these emergences will consequently need to increase dramatically over time. This also means that at some future stage, we may find ourselves left behind as new forms of intelligence develop that are far beyond our understanding or ability to compete with.

These intelligences will not simply be faster versions of human intelligence. Just as self-awareness, complex language, introspection, and consciousness are incomprehensible concepts for a chimpanzee to contemplate (though the chimp may actually exhibit some degree of these qualities or abilities), a radical future emergence could very possibly be beyond our current ability to conceive or comprehend. Such an intelligence won't necessarily emulate all aspects of human intelligence, though it may exhibit and benefit from some. However, it will almost certainly be highly foreign to us, and from its perspective, we may seem very primitive indeed.

The trend toward emergent intelligence that we have traced in this chapter has enormous consequences. If the universe self-organizes at the edge of chaos to generate emergences that maximize entropy production,[17] then that trend should continue throughout the life of the cosmos. Because of this, self-organized complexity and the emergent processes that result from it will probably continue to increase their demand for energy exponentially.[18] As we will see in chapter 18, this trend could be taken to serious extremes in the future.

Based on these processes, we should expect one day to see still greater complexity and new emergences on our own planet, possibly including the eventual appearance of technological superintelligences. Elsewhere in the universe, with its two trillion (2×10^{12}) galaxies and perhaps seven billion trillion (7×10^{23}) stars,[19] similar but different scenarios are possibly taking place. While the influence of each such civilization on the cosmos should be miniscule, as an ever-increasing phenomenon, their collective impact could be significant enough that it might shorten the useful life span of the universe. At the same time, we need to anticipate a number of increasingly powerful emergent intelligences here on Earth and eventually in our little pocket of the cosmos. What those future intelligences will be like is anybody's guess, but there is no reason to think that humanity is the end of the road in this nearly fourteen-billion-year process. In fact, our planet is already experiencing a new emergent phenomenon that has a far higher level of complexity and energy rate density than *Homo sapiens*, as we will see in the next chapter.

CHAPTER 6

THE COEVOLUTION OF HUMANITY AND TECHNOLOGY (PART 1)

"Our technology, our machines, is part of our humanity. We created them to extend ourselves, and that is what is unique about human beings."
—Ray Kurzweil, inventor, futurist, author

"Humans are the reproductive organs of technology."
—Kevin Kelly, editor, author, techno-visionary

Before we time-jump nearly all the way back to the present day, let's first pay a quick visit to our hominid ancestors from about 3.4 million years ago. *Australopithecus* was the immediate predecessor to genus *Homo* and roamed an extended range of north and east Africa for some two and a half million years. Its fossilized remains have been found throughout the region, with many discovered in the part of the Great Rift Valley that extends from present-day Ethiopia to Tanzania. It was here, in the Dikika area of the Afar region of Ethiopia, that the oldest evidence of tool use by our early hominid ancestors was recently discovered.[1] Dated at 3.4 million years ago, these tools represent humankind's earliest technology.

The creation and use of edged stone tools was a major milestone in early human history because it set us on a path that was to change our species forever. We are hardly the first animal to use or even to fashion tools. Crows, orangutans, elephants, and even rodents have been known to manipulate twigs, sticks, and stones as tools. But human beings are the first, and so far only, species to routinely transform natural resources into tools and machines. In doing so, we ourselves have been transformed.

Ranging in height between 3'11" and 4'7" as adults, *Australopithecus* had a brain that was about a third of the size of a modern human's. It is likely they had an intellect only a little greater than that of a chimpanzee. For this species to have discovered how to put edge to stone consistently and methodically and pass that knowledge on to later generations is astounding.

The making of edged stone tools is far more demanding than many people realize, and it is safe to assume it would have been that much more challenging for our diminutive ancestors. Anthropologist Dietrich Stout has been researching this process, known as stone knapping, and its influence on human evolution for many years.[2] The process of putting an edge on stone takes considerable attention, patience, strength, and fine eye-hand coordination. But perhaps most importantly, it takes a special ability to recall and follow step-by-step procedures, keeping each step in its proper sequence. While these are things we now do with relative ease, for *Australopithecus* they would have been hard-won skills. The painstaking process had to be learned, retained, and passed along, all by a species that was not only preliterate, but for all intents and purposes preverbal as well.

Stone knapping involves the incremental removal of material from a piece of fracturable rock, such as flint, by striking it firmly with a harder hammer stone. As the flint is struck repeatedly with controlled force at a precise location and angle, tiny bits of the stone fracture and fall away from the main piece. Considerable care has to be taken, because one slip-up is all it takes to ruin a tool completely. For our ancestors, this was costly not only in terms of time, but also because there was only a limited amount of source material in any given region. This may not have been a problem over a single lifetime, but across 150,000 generations it becomes a motivation for conservation and recycling,[3] as well as for relocation many times over.

In his research, Stout found that a modern human needs more than a hundred hours of practice to become just moderately proficient at stone knapping. We can imagine the challenges it created for the earliest pioneers of this technology. Obviously, though, it was well worth the effort. Not only did these tools make it easier for *Australopithecus* and later humans to slaughter game, they also made it possible to scrape every last morsel of meat from their quarry's carcass. When the hominid was faced with predators and rival clans, the tools also made formidable weapons for protection.

Once the technique for making stone tools was discovered, it would have been no mean feat to consistently pass that knowledge along, because our ancestors didn't communicate as we do. According to paleoanthropologists, it would be nearly three million years before some hominid species became capable of complex speech as we know it. A necessary shift of the hyoid bone, located in the larynx, which is critical in forming the complex sounds of modern speech, didn't occur until about 300,000 years ago. For years, it was believed that only the Neanderthals and *Homo sapiens* experienced this shift in the hyoid's location. Without this shift, early hominids would have been limited to grunts, squeaks, and cries, much the same as in chimpanzees and other animals. While some would classify such animal noises as language, they are far from the equivalent of human speech. Animal calls are nearly always declarative in nature with little faculty for delineating past, present, and future, for abstract concepts, or even for negation.[4] They aren't made up of discrete units that can be recombined in order to alter meaning, as human language does. Additionally, animals don't have a mutation in the FOXP2 gene that is considered critical for the proper development of speech and language, at least in modern humans.[5] It is estimated that this modern mutation didn't occur until some 200,000 years ago.

Stone knapping isn't just the world's oldest technology; in many respects, it is also its most successful one as well. Beginning well over three million years ago, stone knapping continued to be used throughout the world nearly up to the present era. Much has been written about how these tools—along with the later harnessing of fire—altered our diet, providing the building blocks for our evolving bodies and brains. But they changed us in other ways, too, because the act of learning this stone-age toolmaking literally *rewired* our brains. In his research, Stout worked with neuroscientists to perform PET, MRI, and DTI[6] scans of the brains of volunteers before and after they had spent the many hours it takes to become just modestly proficient at knapping edges onto stone. Stout's team found that in the process of learning this skill, specific areas of the volunteers' brains underwent significant changes, most notably in the many new connections that propagated between neurons. The changes included an increase in the number of new dendrites linking neurons in the inferior frontal gyrus (IFG), a region associated with fine motor skills, sequence prediction, understanding complex manual actions, impulse control, and attention. This finding agrees with many of the current models of learning and neurogenesis: learning new tasks, especially over an extended period of time, contributes to changes in the way neurons are linked. Additionally, because the process of creating these tools took place over so many generations and enhanced our ability to survive, Stout surmises that stone knapping would have been a significant evolutionary driver. Any genetic mutation or epigenetic development that increased an individual's ability to learn or transmit this knowledge more easily would have been selected for according to the processes of natural selection. Whether or not this process could have also contributed to greater socialization as well as an increased ability to mirror and model external behavior and internal states remains speculation. In light of the rapid development of the frontal cortex over only a few million years, it seems likely stone knapping was one of many contributing factors that drove our brain's evolution.

But there's something else quite marvelous about the evolutionary development of the inferior frontal gyrus because of how it was later co-opted—or *exapted*[7]—for an entirely different purpose. While the IFG on one side of the brain continues to perform its original functions, in modern humans the IFG in the dominant (typically left) hemisphere corresponds to Broca's area (Brodmann areas 44 and 45). Broca's area was identified in the mid-nineteenth century when it was discovered to be critical to speech production. Damage to this region can cause aphasia, which impedes speech. Considering how the IFG evolved to deal with fine motor control, sequential procedures, and attention, its *exaptation* for speech should not be entirely surprising in hindsight. We may assume that this capability provided greater survivability for the increasingly socially communicative human species.

Toolmaking led to a uniquely virtuous spiral. Once humanity began to manufacture and use tools, over the next two million years or so, our predecessor's brains nearly tripled in size, which along with some other developments paved the way for another major milestone in human history: language.

Though we often talk about animals having their own forms of language, what they really have is a means of signaling to each other. In this sense, animal communication is more similar to the signaling performed by cells, in that the messaging is for the most part fixed, declarative, and its meaning cannot be significantly altered through recombination. Predominantly determined by a species' genes, animal sounds will vary somewhat according to cohort and locale, but these are little more than variations in dialect.

Human language, on the other hand, has grown to be exceptionally rich, diverse, and adaptable. It had to be if it was to create the new vocabularies needed to conceive and describe the ever-developing forms and functions of our technological world. Human language can indicate an immediate situation, or it can discuss something that has occurred in the past or will happen in the future. It allows us to talk about the most abstract, insubstantial of concepts, from sympathy to quantum mechanics. Likewise, negation, irony, humor, and double-entendres are all unique to human language.

Of course, language itself did not suddenly appear fully formed, capable of all of the detail and nuance it has today. It too would have come into being over a great many generations. This shouldn't surprise us, since language is itself a technology. It's just that in this case, instead of manipulating sticks and stones or atoms and molecules in order to alter our environment, language manipulates symbols in order to represent and transform concepts about our world. Simple concepts at the outset, to be sure, but concepts all the same.

The repositioning of the hyoid bone that contributed to the capacity for complex language was once thought to have occurred only in Neanderthals and *Homo sapiens*. But given how unlikely it is that this mutation would have occurred independently in two separate hominid branches, it seems far more likely that it was a trait shared by our last common ancestor, *Homo heidelbergensis*. This particular ancestor lived between 1.7 million and 200,000 years ago. Recently, researchers have found evidence of a similarly located hyoid in this species.[8] It also appears that *H. heidelbergensis* may have been the first species to manage the controlled use of fire and to have customs, including burial rituals.

According to techno-philosopher Kevin Kelly, language, customs, and fire are all forms of technology, part of the *technium*, as he puts it. These three things would have created a positive feedback loop that contributed to the rise of culture for each of these species: language facilitated and was reinforced by conceptual thinking, technologies and customs resulted from and fed back into that language, and the controlled use of fire altered our diet and created a central hearth around which a society of shared knowledge and experience could be born.

The neuroplasticity of the human brain has afforded us a unique opportunity on this planet. In biology, each organism performs a role or function that allows it to occupy a particular opportunity—or *niche*—in the ecosystem. According to a hypothesis called *cognitive niche theory*,[9] instead of evolving in response to environmental conditions and threats, as other species do, human beings incrementally developed the ability to use abstract thinking, reasoning, planning, and cooperation to overcome both our predators and our prey, allowing us to survive and eat better. Where other species had to evolve in response to competitors and environmental change, cognitive tools allowed humans to adapt far more rapidly, offering what would in time become an overwhelming competitive advantage. This would have been a lengthy, but virtuous cycle that promoted not only our survival but the developing cultural and technological basis to eventually overwhelm all competition as we came to dominate the global ecosystem.

Cognitive psychologist Steven Pinker notes that these changes wouldn't have happened overnight. "I think it was a gradual process, reflecting increasing dependence on products of cognition versus instinct, so there is no discrete date at which it began. Certainly, the first use of tools would be an indication that the process had started." According Pinker, "by developing mental models of the environment, cause and effect, the texture of the world around us, and manipulating it to our advantage,"[10] we were able to create and fill an ecological niche in which we are unrivaled. This process would have been self-reinforcing, promoting language, abstract thinking, and socialization.

Occupying this niche contributed to the development of a collective intelligence unlike anything our species had previously exhibited and unlike anything our planet had ever seen. Which brings us back to a critical component of emergence: there needs to be a level of collective interaction between constituents in order for a new emergent behavior to arise. In ants, termites, and swallows, a mere handful of instinctive rules can set the stage for a new emergence. But because cognition works on entirely different timescales than naturally selected instinct, such rules can't provide the needed level of intercommunication. Speech is able to do this, however. Where chemistry provided the means for collective intelligence in cells and instinct in earlier forms of multicellular organisms, language was the method by which humans could extend individual acts of cognition and the behaviors that derive from it. Language allowed us to transcend our own minds as a means of interacting with our peers and, eventually, potentially all of those minds that are downstream from us.

As we now know, *H. heidelbergensis* and the Neanderthals came to their ends about 200,000 and 40,000 years ago, respectively. But their legacy of tool use and culture was carried on by our own species, *Homo sapiens*. As has been detailed in many other places, the human race continued to incrementally advance, building culture and shared knowledge across many millennia. The First Agricultural Revolution,[11] approximately 12,000 years ago, gave us roots, slowly transforming us from nomadic clans into anchored communities that were more closely linked to the land and therefore to a specific location. Social bonds grew, contributing to the power of numbers, shared knowledge, division of labor, and specialization. Writing transformed knowledge from something that could be transmitted only through immediate interaction into transportable information capable of transcending time and place. Civilizations grew and fell, slowly building knowledge that proliferated, was occasionally lost, and was frequently rediscovered. Centers of learning arose. Renaissances blossomed. Science and learning were institutionalized. And now, here we are.

Meanwhile, all this time another trend was building. As knowledge and technology built upon their past successes, something wondrous was taking place. Little by little, all of it—knowledge, technology, culture—was accreting, adding to itself, accelerating the progress that had previously been made. This phenomenon of accelerating change would eventually become well recognized. Perspectives like those of *Future Shock* author Alvin Toffler eventually transformed the concept of accelerating progress into a broader conversation. In fact, the pace of change was accelerating in a consistent pattern when viewed over the decades, the centuries, the millennia. This is not to say that the rate itself was necessarily changing, but because it is exponential in nature, it eventually catches up to and advances far more rapidly than linear systems, eventually surpassing human time frames and life spans. As a result, what once took eons, then centuries, then years, eventually took mere days and will eventually take only hours. In time, the pace of change becomes biologically impossible to adapt to and can only be dealt with through the use of more technology.

Which brings us back to the coevolution of humanity and technology. For three and a half million years, we have been altered by our technology as we have been altering it. We have taken technology from stones and simple machines to particle accelerators and quantum computing. At the same time, technology has taken us from preliterate, preverbal primates to a speaking, socially interdependent hominid to a globe-spanning species on the verge of interplanetary travel.

During this time, we have collectivized our brain power and knowledge. Our collective intelligence doesn't just result in new and better ways of doing things. It actually manifests technology, society, and civilization, which are so much more than the sum of us. This interplay of ideas and economics is in a constant flux between the needs of the individual and the needs of the collective, of society, much as occurs in the cells in an organism.

As with previous emergences, society processes energy at a far faster rate than the individual configurations that came before it. That is, individual humans. Looking back on Chaisson's calculations of energy rate density, we see that modern society processes energy twelve times faster—more than an order of magnitude—than individual animals do, including our hominid ancestors and ourselves.

Moreover, this is not a fixed value. Civilization has only leapt to this level in the past handful of centuries, having climbed incrementally, but exponentially, over the thousands and millions of years before that. There is little to suggest that the pace of increase is going to stop.

But it is also evident that something will have to change. Just as we would burn up were we to try to continuously process twelve times as much energy through our fragile bodies, society's overall energy processing and consumption is having a devastating impact on our planet. The day is fast approaching when we will have to cease dumping all of our excess waste energy into Earth's contained ecosystem and find ways of radiating it into the cosmic heatsink of the greater universe in some controlled manner. This will be essential if we are to survive the relentless trend of energy consumption.

Humanity and technology have been converging, making us ever more reliant on our machines. From the first crude tools that were fashioned to fit in our hands to wearable computers that encircle our wrists, our tools have increasingly become a part of us. This is a trend that has gone on for three and a half million years and is all but certain to continue for the rest of our existence as a species. From edged stone tools to artificial limbs, contact lenses, and cochlear implants, we have designed and transformed technology to do our bidding, to shape our world, to make our lives better and easier. At the same time, technology has changed us as well. But while it has certainly shaped our bodies, it is our brains and the way we think that have seen the greatest transformations.

Leaving our hominid relatives behind, it is time we returned to the present day. Almost.

The date is July 10, 1956, as we set down on the Dartmouth College Green, our ship's chameleon-mode allowing us to remain hidden as students come and go about their day. Nestled along the Connecticut River, in the town of Hanover, New Hampshire, Dartmouth is one of the oldest colleges in the United States, established well before the American Revolutionary War.

Insect-like camera drones depart our craft and flutter over the Green, heading for the top floor of Dartmouth Hall, where a small group of men sit listening to a lecture. The lecturer, Ray Solomonoff , is thin, with a ring of dark hair that encircles his shiny pate. Some of the attendees take notes as he discusses the use of inductive inference as a means of making computers improve themselves. Needless to say, this is not a typical topic for any institution, here in the middle of the twentieth century.

We are here to witness a milestone that by our present day has assumed mythic proportions, for this is when a small band of carefully

selected scientists assembled for a two-month workshop that was
to launch the field of artificial intelligence, setting technology on a
path that would change our world forever. Running from June 17 to
August 18 and formally known as the "Dartmouth Summer Research
Project on Artificial Intelligence," the workshop raised very high
hopes. According to the proposal authored by John McCarthy, Marvin
Minsky, Nathaniel Rochester, and Claude Shannon, "an attempt will
be made to find how to make machines use language, form abstractions and concepts, solve kinds of problems now reserved for humans,
and improve themselves. We think that a significant advance can be
made in one or more of these problems if a carefully selected group
of scientists work on it together for a summer."

   We now know that these scientists had vastly underestimated the
challenges they had set for themselves. Even though the organizers
and attendees would become legends over the coming decades, there
was simply so much they didn't know about the problems they faced.
The first programmable fully automatic digital computer, the Z3, had
only been invented fifteen years earlier, and the first computer to
use a true von Neumann stored-program architecture, Manchester
Baby, wasn't built until 1948. Though major accomplishments in their
day, the early computers of the mid-twentieth century were a far cry
in flexibility, abstraction, memory, and processing power from what
would be needed to meet many of the workshop's goals.

   The term *artificial intelligence* was actually first coined for this workshop by organizer John McCarthy, who wanted what he thought
would be a neutral term for the group. This was because so many of
these scientists, some of them still in their twenties, already had their
own areas of specialization and focus, and therefore their own agendas
as well. McCarthy, who was working at Dartmouth at the time, hoped
to solidify the orientation of this nascent field and explore the possibilities of computer intelligence. He would later go on to invent LISP,
one of the earliest high-level programming languages that would be
widely used in artificial intelligence research. Co-organizer Marvin
Minsky, who would go on to found MIT's Artificial Intelligence

Laboratory, had already been working on neural network learning machines since 1951. Claude Shannon, who is widely considered the father of information theory, sought to apply his work to computers as a means of overcoming certain intelligence hurdles. Ray Solomonoff, whose prodigious note-taking documented much of the project, was focused on developing algorithmic information theory and finding a way to use probability to make computers more generally intelligent. Likewise, the workshop's other participants each had their own specific areas of expertise and focus.[12]

While the Dartmouth workshop fell short of its Herculean goals, it succeeded in being an intense and extended sharing of ideas at this critical moment in computer history. Many of the participants credited its cross-pollination of ideas and approaches with influencing their own work in the years that followed. This was particularly true for symbolic information processing, which was perhaps best represented at Dartmouth by the work of Allen Newell and Herbert Simon. With the help of programmer Clifford Shaw, their Logic Theorist software was completed that summer and is considered the first artificial intelligence program. Designed to prove theorems, Logic Theorist proved thirty-eight of the first fifty-two theorems in Whitehead and Russell's *Principia Mathematica*,[13] some more elegantly than the authors themselves. Newell, Simon, and Shaw's work would go on to influence the development of many early AI programs, including the team's own General Problem Solver, created in 1959.

Thus began a recurring debate in the field of AI, among those who believed success lay in using symbolic logic and those who believed a connectionist approach was needed. The symbolic camp looked to the long history of philosophy and mathematics, including the work of Gottfried Leibniz and George Boole. A mathematician, logician, and philosopher, Boole is perhaps best known for his development of Boolean algebra, the logic that is at the heart of most modern computer programs.[14] Leibniz was one of history's great logicians, a polymath and philosopher who invented differential and integral calculus independently of Isaac Newton. One of Leibniz's goals was the creation of a *characteristica universalis*,[15] a universal language of logic and reason.

The connectionists took a very different approach to achieving artificial intelligence, influenced in part by relatively recent ideas based on our growing understanding of biological brains and neural circuits. However, there were several factors that would limit what could be achieved with artificial neural networks for the next few decades, with the result that symbolic logic was the predominant approach in the race to develop AI during that initial period.

In the years that followed, the field of artificial intelligence would meet with many early successes. The developing ability of computers to solve problems, prove theorems, use language, and play games convinced many researchers and agencies, as well as journalists and the public, that human-level machine intelligence was just around the corner—or if not, then at least that it would be achieved in a matter of only a few decades.

One popular early method, known as *reasoning as search*, approached goals by proceeding through a problem in a step-by-step manner. This could work well for very simple problems, but as the challenges became more complex, this method routinely came up against a "combinatorial explosion" of possibilities. While it soon became evident that this precluded reasoning as search being used for more complex problems, the issue also eventually led to the development of heuristics in artificial intelligence, a general technique for using imperfect but more practical methods of approximating and determining an optimal solution to a problem in a manageable time frame.

Artificial intelligence and related fields such as robotics realized many other advances in those early years. Programs and techniques were developed that allowed computers to play checkers at a respectable amateur level. Minsky developed a system that used a robot arm, a video camera, and a computer to assemble children's building blocks. Natural language processing (NLP) had some early successes, allowing systems to recognize and translate from one language to another. One NLP program was ELIZA, the first chatbot. Invented between

1964 and 1966, ELIZA was directed by scripts. One particular script, DOCTOR, emulated a Rogerian psychotherapist so successfully that many users seemed to forget they were interacting with a software program.

During this period, millions of dollars flowed to researchers in an effort to realize this dream of human-level artificial intelligence. But eventually there was too great a difference between the promise and the payoff. In 1966, the skeptical ALPAC report commissioned by the US government led to a serious reduction of funding for NLP research. Then in 1973, the UK's Lighthill report[16] was particularly critical of the lack of advances made in AI. Government funding subsequently dried up on both sides of the Atlantic, and what later became known as the first AI winter set in.

The first AI winter, lasting from 1974 to 1980, made AI research harder, but progress continued nevertheless. By 1980, a new form of AI was gaining ground. *Expert systems* were an effort to replicate the decision processes of human specialists, such as in medical diagnostics, tax advice, and risk assessment. These systems were generally made up of a knowledge base drawing from human expertise and an inference engine that applied these rules according to conditions and branching logic. While many of the systems showed promise early on, their brittleness based on the inputted data soon became evident and led to knowledge bases eventually falling out of favor by around 1987. This echoed the pattern seen leading up to the first AI winter, when the hype surrounding expectations far exceeded the realities of what could actually be delivered. A second AI winter ensued, a period that historians would later say lasted from 1987 to 1993.

Work continued to progress in artificial intelligence, despite the need for occasional readjustments of expectation on the part of the government, industry, and the public. One area of AI that made a resurgence was artificial neural networks, or ANNs. Early networks, including single-layer neural networks known as perceptrons,[17] fell out of favor following Minsky and Seymour Papert's criticism of them in their 1969 book *Perceptrons*.[18] In it, they discussed the major shortcomings of this approach: its inability to process *exclusive or* logic operations—otherwise known as XOR—and the lack of necessary processing power required for a usefully large multilayered neural network.

Then in 1974, Paul Werbos invented the backpropagation algorithm that resolved the XOR problem,[19] making the training of multilayered neural networks feasible. Still, sufficient processing power remained a major challenge, and alternate machine learning methods such as *support vector machines* grew in popularity from the late 1980s and early '90s.

In the mid-2000s, a number of things happened that made neural networks far more capable and powerful, bringing them back into favor. The first was the ongoing trend of Moore's law, the regular doubling of the number of transistors and other components that could fit on a microprocessor. In the thirty years between 1975 and 2005, the transistor count on a high-end CPU had increased by a factor of about 100,000, which represents a doubling approximately every year and a half. This made vastly more powerful computers possible and is a prime example of the power of exponential growth. Such dramatic improvement made it possible to run far more complex software, including, eventually, neural networks.

By itself, the increase in processing power still wasn't enough to overcome the other challenges of many layered neural networks. However, around that time there were a number of key software advances, including a pair of pivotal papers coauthored by University of Toronto computer scientist Geoffrey Hinton and Ruslan Salakhutdinov.[20] Their papers described how to make a considerable portion of the process of setting up a neural network unsupervised. Traditionally, much of the setup of these networks had to be carefully supervised and balanced, with developers painstakingly involved in adjusting the weighting and other parameters of each node and layer. But the techniques presented in these papers eventually made it possible to not just set up a network more quickly but to use far more layers, which was essential for many complex tasks.

The third factor promoting the ascendance of neural networks around that time was an increase in the availability of and substantial price drop in GPUs. Graphic processing units were originally designed to speed up image processing for everything from computer gaming graphics to 3-D image displays. Because of this, they are optimized for certain types of highly parallel mathematical calculations, such as matrix and vector operations. As it happens, the same operations can also benefit neural network processing. These and other advances led to far more complex neural networks than had ever been possible before.

The resulting improvements in AI led to the popularization of the term "deep learning," which had first been introduced to the machine learning community in 1986 and made manifest by the first convolutional neural networks, including LeNet–5 developed by Yann LeCun in 1998. (Hinton, LeCun, and University of Montreal professor Yoshua Bengio were awarded the one-million-dollar 2018 Turing Award by the Association of Computing Machinery—sometimes referred to as the Nobel Prize of computing—for their contributions to AI and deep learning.)

It seems that through the years, AI has been plagued with unrealistic expectations regarding its rate of advance. Nevertheless, improvements during the past decade seem to have come at a whirlwind pace. How rapidly did these technologies advance? By 2020, several metrics indicated progress in machine learning was doubling every eighteen months. Even more astoundingly, the compute (the number of processing cycles) used to train large AI models was doubling every 3.43 months.

To put this in perspective, in 2017 DeepMind's Go-playing computer AlphaGo Zero[21] ran 300,000 times as much compute as AlexNet, the convolutional neural network that soundly won the ImageNet Challenge in 2012.[22] All of this additional processing allowed AlphaGo Zero to master the complexity of Go without using data from any human matches. Instead, it trained for three days by playing 4.9 million games against itself, learning the game as it went, eventually competing at superhuman levels.

Much of this improvement in processing power was due to the transition from GPUs to tensor processing units, or TPUs. Where GPUs had been an off-the-shelf method of improving neural network performance a decade earlier, TPUs were application-specific integrated circuits (ASICs) developed by Google specifically for the task.

Today there is a veritable menagerie of neural networks that are used for different types of tasks. Convolutional neural networks (CNNs) are widely used in image recognition and natural language processing. Recurrent neural networks (RNNs) are better optimized for dealing with processes that involve time series data, making them better suited for robotic control, speech and gesture recognition, and machine translation. Long Short-Term Memory (LSTM) is a form of RNN that processes temporal data with long-range dependencies, leading to much greater accuracy in applications such as speech recognition and sign language translation. In all, there are dozens of flavors of different neural networks, with some of the earliest forms long having fallen out of favor.

A rapidly increasing proportion of our world and our lives are affected by artificial intelligence. The cars we drive, the social media we share, the hotel and flight reservation systems we book on, the fraud detection systems that flag our credit cards when an unusual purchase is made—all of these advances are transforming our world and the ways we interact with it, bringing greater convenience, but new challenges as well.

An interesting aspect of our attitudes to AI is that as each new development in artificial intelligence is made and incorporated into our lives, we tend to stop thinking of it as artificial intelligence. In fact, we tend to stop thinking about it at all. Optical character recognition, recommendation systems, in-camera face detection, voice-activated smartphones, and email spam filters all use different forms of artificial intelligence, but once they reach the point that they work dependably, we seem to lose sight of this, allowing them to fade into the background of our lives. As a result, for many of us, artificial intelligence seems to be continually just around the corner. Except the reality is that it is here now and relentlessly becoming ever more pervasive. How far this will go is anyone's guess, but from a standpoint of physics and the laws of nature, no endpoint appears imminent. The process of increasingly more intelligent and responsive technology could continue to transform our world for a very, very long time.

What happens once technology reaches some as yet unknown threshold? When it not only exceeds human intelligence, if that is possible, but becomes capable of rewriting its own code and restructuring its own hardware in order to improve itself? The latter threshold of implementing methods of self-improvement has already been crossed to some degree and will likely be pushed much further in coming years and decades. Self-modifying software, dynamic reconfiguration using field programmable gate arrays (FPGAs), auto-machine learning, and other approaches that will be explored later have all passed this threshold of AI self-improvement.

Hominids embraced technology long, long ago and never looked back. That relationship transformed us and raised us up to such a degree that we barely recognize ourselves in our early progenitors. In the process, we have developed and advanced technology as well, taking it from the crudest of reshaped stones to the reconfiguration of molecules that extend our senses, minds, and bodies. With our tools, we now view the farthest reaches of the cosmos, manipulate energy and matter at subatomic levels, and routinely gain new insights into previously hidden secrets of the universe.

Over the course of human evolution, so much has been accelerating. Technological change, cultural change, even evolution itself has been speeding up.[23] Consider that our Australopithecine ancestors entered into this revolutionary relationship with technology three and a half million years ago. *Homo sapiens* appeared after 90 percent of that time (around 300,000 years ago), and the first civilizations after perhaps 98 percent of the remaining time span. This should give us lots of perspective. Put another way, the period since the beginning of the Scientific Revolution, which began around 1500 AD, represents 0.02

percent of our long ascent with technology—in terms of time, that is. In terms of knowledge, inventions, and cultural advancement, the relationship is practically the inverse. Far and away the majority of all human and technological progress has been accomplished during that relative blink of an eye.

Let us consider Earth's 4.55 billion years compressed into a single calendar year, with our planet forming on January 1, a concept popularized by cosmologist Carl Sagan. While life first appeared on January 28 (or February 25, if we use more conservative estimates), it wouldn't be until November 12 that the first multicellular organisms evolved, with hominids arriving on the scene at about noon on December 31. *Homo sapiens* appeared at 11:36 p.m. on that final day, and the Scientific Revolution (1543 AD) began after 11:59:56 p.m. Finally, the first digital computers were invented around half a second before midnight on the last day of this imaginary calendar year.

Just think about how long it took to go from single-celled life to multicellular organisms to brains to *Homo sapiens*. Now compare that to the time it took to get to where technology is today, whether we set our starting point at the origins of stone toolmaking, the Scientific Revolution, or the birth of the computer. From this perspective, human-level technological intelligence and beyond is really just around the corner, whether that takes three decades, three centuries, or three millennia. I'm betting it is a lot closer to the nearer end of that range. Which brings us to our next question: What wonders can we expect to witness during the next chapter of coevolution between humanity and technology?

TWENTY-FIRST
CENTURY

CHAPTER 7

CHATBOTS AND THE ILLUSION
OF AWARENESS

"The future of computer power is pure simplicity."
—Douglas Adams, author of *The Hitchhiker's Guide to the Galaxy*

"Alexa, let's chat."

It's November 24, 2017, at Amazon's Day 1 headquarters in Seattle. I follow my escort through security and upstairs to the fifth floor, where I'm ushered into a room partitioned with walls of floor-to-ceiling black curtains. Brushing past the curtains, I take my seat in a modern wingback chair at the center of the room. On a small table is a writing tablet, a pen, and a pair of Sony studio headphones, as well as a matchbox-sized device that has a single button. At my left, a round silver reflector bounces illumination from the blindingly bright studio light set in front of me. In the darkness beyond that, I can just make out the figure of my videographer perched behind his tripod and digital camera, waiting patiently for our session to begin. I don the headphones and we run through a quick sound check. Though I can't see them, I know there are two other judges who, like me, are sitting in similar ad hoc mini studios.

The voice of our interactor comes through my headphones loud and clear as he resumes this round of the finals with the trigger phrase: "Alexa, let's chat." This is immediately followed by the now familiar computer-generated voice of Alexa, the virtual digital assistant software that runs on an ever-growing number of Amazon-supported appliances.

"Welcome to the Alexa Prize Grand Finale. Here's one of the finalists. Hello! This is an Alexa Prize socialbot. I think I have already heard your voice. Have we talked before?"

"We have," replies the interactor, one of three people selected to engage the competing socialbots in conversation according to specific guidelines.

"What a faithful chatmate you are. What was your name again?"

"Mike."

"Yeah, I think I remember you, Mike. But let's talk about you. How is your day going? Would you share with me some of your today's terrific experiences?"

We are in the midst of the first Alexa Prize, Amazon's $2.5 million competition designed to spur innovation in the field of conversational AI technology. Alexa is a voice service that runs programs known as skills. Each of these team's conversational socialbots is one such skill. By 2019, there were more than 90,000 Alexa skills, from fitness trainers to smart home controllers to a version of the television game show *Jeopardy!* There are now hundreds of third-party devices with Alexa built in, plus Amazon's ever-growing list of devices, including the Amazon Echo on which it was first introduced to the public. More than 60,000 smart home products can be controlled with Alexa, including lighting, ceiling fans, thermostats, and smart TVs.

More than one hundred university teams from twenty-two countries applied to compete in the first year of the Alexa Prize competition, from which fifteen were eventually selected. After working on their socialbots for much of the year, including training them through interactions with the public, the competition is now down to the remaining three finalist teams. In this first round of the finals, three judges, including me, are grading conversations between a human interactor and each team's socialbot. It is the job of the interactor to engage Alexa in what will hopefully be a coherent and interesting conversation. Once a judge feels a socialbot has become too repetitive or nonsensical, they press the small button at their table. After two judges drop the conversation, the conversation is stopped. The goal for each team is to keep the conversation going for as long as they can—ideally at least twenty minutes, the threshold for winning the grand prize. There are three separate rounds, with a different interactor and three different judges in each round. All of the socialbots and teams remain anonymous to the judges and interactors throughout the finals.

Though all of the socialbots speak with the same computer-generated voice, each team is running very different software. The programs sail through or stumble over different things, but overall, they are all surprisingly capable. In the end, Sounding Board, a team from the University of Washington, is the winner of a $500,000 prize to be split among themselves. Their team is made up of five doctoral students with expertise in language generation, deep learning, reinforcement learning, psychologically infused natural language processing, and human-AI collaboration, with guidance from three electrical engineering and computer science professors. Sounding Board managed to maintain an engaging conversation for an average of ten minutes and twenty-two seconds, shy of the twenty-minute threshold, though one of its conversations came very close. Had their socialbot managed to surpass this hurdle with two of the interactors and achieved an average rating from the judges of 4.0, they would have won an additional $1 million for their university.

Alexa is far from the only digital assistant in the world. All of the giants of AI are developing their own unique version of this potentially transformative interface. Microsoft's Cortana has been a presence on the Windows platform since 2015, beginning with Windows 10 and expanding to numerous other devices. Apple's Siri was originally released in 2011 on the iPhone 4S, and was subsequently made available on all of their iOS devices.

Drawing on their immense search engine infrastructure, Google developed Assistant, which was released in 2016. Google Assistant does a good job answering questions and performing tasks, responding quickly to spoken requests, but as with its competitors, it still has some way to go.

In 2018, Google Assistant became more accomplished with the integration of Google Duplex,[1] an extension that allowed Assistant to autonomously place a call and schedule appointments and reservations with a person at the other end of the line. For the most part, the voice that Duplex uses is indistinguishable from a human voice, just one of the features that wowed listeners when it was first demonstrated in May 2018.

Woman: Hello, how can I help you?

Duplex: Hi, I'm calling to book a woman's haircut for a client? Umm, I'm looking for something on May third.

Woman: Sure, give me one second.... Sure, what time are you looking for around?

Duplex: At 12 p.m.

Woman: We do not have anything at 12 p.m. available. The closest we have to that is a 1:15.

Duplex: Do you have anything between 10 a.m. and uhh, 12 p.m.?

Woman: Depending on what service she would like. What service is she looking for?

Duplex: Just a woman's haircut for now.

Woman: Okay, we have a 10 o'clock.

Duplex: 10:00 a.m. is fine.

Woman: Okay, what's her first name?

Duplex: The first name is Lisa.

Woman: Okay, perfect. So, I will see Lisa at 10 o'clock on May third.

Duplex: Okay, great. Thanks.

Woman: Great. Have a great day. Bye.

People were amazed not only at the ability of the program to nego-tiate the complexities of human conversation, but also how lifelike the speech was. This was realistic and nuanced, right down to the cadence, pauses, and filler words, such as "umm" and "uh." However, the demo raised serious questions almost immediately. The lack of identification of the business or person answering made some people wonder if the demo was canned or faked. For others, the fact that Duplex sounded so much like a human, yet hadn't identified itself as a bot, was of far greater concern, given the ethical issues it raised as well as the potential for abuse. Responding to this, the next month Google gave a second, lower-key demo that addressed most of these issues, especially the matter of identifying itself as "Google's auto-mated booking service" at the beginning of the call. As impressive as the program was, at the time of the demos, Duplex could only make haircut appointments and restaurant reservations and answer inqui-ries about business hours. Additionally, subsequent reports indicated portions of these exchanges were passed over to a human operator at some stage, presumably to further train the system. Nevertheless, Google clearly demonstrated the direction they want to take this technology.

Not to be outdone, in China, AI giant Baidu has developed DuerOS, a conversational AI system for use on devices, including its own Xiaodu and Raven smart speakers. DuerOS intelligent chips can be integrated into virtually any device to provide the functionality of its assistant software. Multinational conglomerate Tencent has devel-oped Xiaowei, a digital assistant that integrates with their WeChat app, a multipurpose messaging, mobile payment, and social media app used by more than a billion people around the world. E-commerce giant Alibaba had enormous incentive to develop its digital assistant, Cainiao, given that the company delivers more than one billion pack-ages a day. While Cainiao is currently being used only for coordinat-ing package deliveries, it will no doubt expand to deal with many different tasks, including the fifty thousand customer service calls the company receives each day.

Clearly, we have reached an inflection point in the development of these digital assistants, but how did we get here? The field of artificial intelligence has a long history with chatbots, which is essentially what these teams are building. The development of these programs goes well beyond a fascination for talking computers. In his 1950 paper, "Computing Machinery and Intelligence,"[2] mathematician, cryptanalyst, and computer scientist Alan Turing laid out his premise for what he called "The Imitation Game," in which a computer attempts to deceive a human judge into thinking it is actually a person. In time, this and other variants described in Turing's paper became known as the Turing test,[3] and computer scientists, researchers, and hobbyists have sought to meet the challenge, either by attempting to create genuinely intelligent programs or through artifice, by using gimmicks to try to fool the human judges.

ELIZA is generally considered the first chatbot program, having been completed in 1966. Developed by MIT AI professor Joseph Weizenbaum, ELIZA was based on a series of scripts that directed the program's responses according to the previous input statement provided by the person it was communicating with. Using its most famous script, DOCTOR, ELIZA would respond much as a Rogerian psychotherapist might. (Rogerian psychotherapy later became more commonly known as person-centered therapy.) For instance, a typical dialogue might go something like:

ELIZA: Is something troubling you?
Person: Men are all alike.
ELIZA: What is the connection, do you suppose?
Person: They're always bugging us about something or other.
ELIZA: Can you think of a specific example?
Person: Well, my boyfriend made me come here.
ELIZA: Is it important to you that your boyfriend made you come here?
Person: He says I'm depressed much of the time.
ELIZA: I am sorry to hear that you're depressed.

Person: It's true. I am unhappy.

ELIZA: Can you explain what made you unhappy?

From our vantage point, this feels like a relatively logical if somewhat stilted conversation, but to a programmer it is a fairly straightforward script, built around a rules-based engine. The statements produced by the computer work by using pattern matching and substitution, incorporating a series of cue words or phrases extracted from the user's prior response.

Ironically, Weizenbaum created the program to demonstrate what he felt was the superficiality of computer communications, but he was surprised by the response of many of the people who engaged with it. Weizenbaum even relates the story of his secretary conversing with the program. Following a few exchanges, she asked Weizenbaum to leave the room so she could be alone with the computer! As he wrote in 1976, "I had not realized . . . that extremely short exposures to a relatively simple computer program could induce powerful delusional thinking in quite normal people."[4] Needless to say, Weizenbaum was successful, though not in the way he had originally intended.

In the years that followed, chatbot technology evolved, and while some developers created new strategies, many continued to build on Weizenbaum's methods, which had been so successful in engaging and fooling human judges.

Unfortunately, the ability to fool a human is no true gauge of machine intelligence. As Weizenbaum unintentionally demonstrated, human-machine interaction is a two-way street, and it is evident we humans are all too ready to ascribe awareness and even personalities to aspects of our environment that have neither. In their book *The Media Equation*,[5] Stanford professors Clifford Nass and Byron Reeves make the argument that we interact with much of our media as though it was another person. This may even extend to an unconscious tendency to treat other technologies—our computers, cars, boats, and other tools—as though they were also alive and self-aware. To quote from their book, "Individuals' interactions with computers, television, and new media are fundamentally social and natural, just like interactions in real life."

If true, this goes a long way toward explaining why, even with the earliest, most basic of chatbots, users have been willing to accept them as social, conversational, and sometimes even intellectual equals. This can even be true for some of those participants who are already aware of the programmed nature of their conversation partner.

Our readiness to accept these programs as conversational partners is a major reason I believe the chatbot and the virtual personal assistant are destined to become *essential* user interfaces in the near future. As technology has advanced, particularly in the field of computing over the last seventy years, the need for well-designed interfaces has become desirable, if not essential. As Brenda Laurel wrote in her 1990 book, *The Art of Human-Computer Interface Design*,[6] "We naturally visualize an interface as the place where contact between two entities occurs. The less alike those two entities are, the more obvious the need for a well-designed interface becomes." In so many respects, our technology is becoming increasingly complex and is in need of a universal, intuitive interface that allows us to easily and rapidly interact with it, whether we are veteran users or engaging the technology for the very first time.

Since the beginning of the computer age, we have sought out increasingly powerful, flexible, and natural user interfaces. Beginning from hardwired programs, we moved to punch cards and punch tape before this gave way to the command line interface, which allowed users to type commands directly into a system. Graphic user interfaces followed, using mice and monitors that could provide a What-You-See-Is-What-You-Get—or WYSIWYG (pronounced wiz-e-wig)—experience. As computers became even more powerful, there was enough spare processing power to perform the work needed to implement a range of natural user interfaces, or NUIs. This progression clearly demonstrates a trend toward ever more natural means of interacting with our increasingly complex technology. Today, NUIs that allow us to use gesture, touch, and voice commands are becoming more and more common as a means of controlling our devices. The overall trend has democratized the use of computers, taking them from devices once only computer scientists could operate, to being accessible to enthusiasts, to today, when they can be used by children and even toddlers. As we enter the third decade of the twenty-first century, language-enabled virtual assistants are being developed to act as our guides in our increasingly computerized world.

A particularly human inclination is driving the development of our progressively social machines. In some respects, this is ironic, because in the past there has been so much concern about our becoming too much like our machines, when in reality we have been making them more and more like us—or, at least, we've given them the ability to operate on our terms as much as possible. Clearly, the move to more intuitive interfaces has hardly run its course. As we'll see in subsequent chapters, organic user interfaces will one day allow us to work even more closely with the technologies we are developing.

In the meantime, let us consider where this new era of virtual assistants is trending and what it is likely to look like in the years ahead. As this technology progresses, exploiting the illusion of awareness and understanding will continue to be seen as a positive feature of the evolving interface. One way this has manifested is in chatbot developers attempting to simulate personalities in their bots. This takes many forms: from bots that attempt to use phrases and idioms in idiosyncratic ways, intending to make them sound more human, to mimicking or mirroring the person they are chatting with. Successfully implemented, this can result in a conversation that can quickly make you forget you're interacting with a software program.

As an example, Mitsuku from Pandorabots engages in surprisingly realistic text-based conversation. Developed by Yorkshire-based Steve Worswick, Mitsuku is a five-time winner of the Loebner Prize, an annual competition to determine the world's most humanlike chatbot. Mitsuku is based on a scripting language called AIML, Artificial Intelligence Markup Language. Structured around categories, patterns, and templates, AIML uses pattern-matching to respond to inputted text messages. AIML was developed by computer scientist Richard Wallace, based initially on his chatbot A.L.I.C.E. (Artificial Linguistic Internet Computer Entity). A.L.I.C.E. was itself inspired by Weizenbaum's ELIZA and sought to extend that program significantly. It should perhaps come as no surprise that A.L.I.C.E., as a direct descendant of ELIZA, became a three-time winner of the Loebner Prize in 2000, 2001, and 2004.

The Loebner Prize is set up much like a fairly typical text-based Turing test, where a chatbot and a hidden human participant each try to convince a human judge that it is the real human. But while the Loebner Prize competition is sometimes equated with the Turing test, many serious AI researchers deride the contest for its lack of rigor. Perhaps this isn't surprising, given that so many researchers in the field think the Turing test itself should be retired. Marvin Minsky himself observed that "the Turing test is a joke. It was suggested by Alan Turing as one way to evaluate a machine, but he never intended it as being the way to decide a machine was really intelligent." While the Turing test has represented an early goalpost for researchers, Minsky is correct. The problem with these tests is that it is too easy for developers to resort to standard chatbot tricks that engage our tendency to anthropomorphize, ascribing personality and awareness to a program that has neither.

It's unlikely the Turing test, both as it was conceived and as it has evolved, will ever be capable of identifying whether or not a machine is truly intelligent. This is due to several factors, not least of which is the very definition of intelligence as explored in chapter 1. It is possible, perhaps even probable, that machines will never be able to fully replicate human intelligence, simply because we start from such very different origins. While shuttling digitized data through transistors can be structured to emulate biological intelligence, such emulation can never be perfect, due to incompatibilities between the two substrates. Of course, it is also possible the digital emulation of biological processes will one day be refined enough that it will be indistinguishable from its human counterpart, but that day is likely to be many decades away. Perhaps much longer.

These considerations aside, advances in chatbot technology fulfill a very important need in the ongoing development of AI. The human proclivity for conversation means there is an endless supply of data for these chatbots to be trained on. Though still far from being the same as speaking with another person, speaking with these chatbots can fulfill a number of very real needs, from battling loneliness to therapy to talking out a strategy or problem.

Given their origin in ELIZA, it should be no surprise that chatbots can be therapeutic in unexpected ways. For example, in late 2015, a Russian software developer and entrepreneur, Roman Mazurenko, was killed when he was struck by a speeding car. Grieving the loss of her best friend, Mazurenko's fellow developer Eugenia Kuyda began building a software program to memorialize him. Training the software on thousands of their archived text message conversations, over time the bot began to respond convincingly like Roman. From this poignant beginning, the software known as Replika was born. Released to the public in 2017, Replika is not focused so much on achieving a kind of digital immortality—now it relies on the user's own input in order to eventually emulate them. Using a deep neural network, the program is trained on a person's conversations with it. It gradually takes on some features of their personality, opinions, and manners of speaking. The program is well suited to free-form conversation, even emulating emotion and empathy to some degree. Occasionally, the results can be surprisingly realistic.

By 2018, the Replika project was made open source under the name CakeChat, allowing developers all over the world to create similarly social chatbots. Using a form of recurrent neural network (RNN) called sequence-to-sequence, these chatbots can be surprisingly emotionally supportive, despite the fact they lack any form of awareness or emotion themselves. This may be because, in many ways, they are mirrors of ourselves, having crudely captured some of the qualities that we identify as human somewhere deep in their neural networks.

To be sure, none of these chatbots is the same as engaging in a conversation with another person, but for some people it seems to provide solace, even companionship. Situations can be imagined where such interaction could be better than none at all. A chatbot that lacks any form of consciousness or empathy may be a poor substitute for a human companion, but for a shut-in or someone in seclusion, it may help maintain their mental health and is probably better than spending time with someone hostile or unpleasant. Several chatbot developers, including those behind Replika and the therapy bot Woebot, have focused on these benefits. A person isolated due to calamitous circumstances—at a polar outpost, on a space mission, or in a Mars colony—would probably be thankful for company, ersatz or otherwise. And while dealing with serious mental illness is well beyond chatbots' current capabilities, existing chatbot software could perform a role in the sort of self-help therapy ordinarily achieved by journaling and self-talk. Similarly, problem-solving is often just a matter of finding the right conduit for framing a challenge, something that might be facilitated by talking with a chatbot as a means of thinking things through.

So, while chatbots still fall short of human-level communication and understanding, including our aptitude for emotional awareness and empathy, not every task requires those human qualities. Between these conversational abilities and the increasing use of speech recognition as a preferred means of controlling and interacting with our technological world, there should be more than enough market incentives to rapidly drive and grow the sophistication of these interfaces.

Let's leave the present day and hop ahead a few years in the first of what will be many excursions into the near and distant future. The year is 2023, and conversational interfaces have become consumers' preferred method of engaging with most of their devices. Driven by the growth of smart home technologies along with advances in natural language processing (NLP), nearly 300 million voice assistant devices are in homes globally.[7] From 25 million devices in 2018, this represents growth of over 1,000 percent over this five-year period. Put another way, this is faster than the sales growth of the iPhone beginning in 2007, which is considered to be one of the most rapidly adopted technologies of all time.

The rapid growth of voice-assisted technology extends to back-end enterprise technologies, where it is becoming an increasingly essential tool for IT departments everywhere. While many tasks are still better suited to graphical user interfaces (GUIs) or even the command line, many others can be performed with authorization via voice recognition and a series of vocal commands—almost like issuing instructions to a trusted human assistant.

Voice technologies will have continued to develop and segment in many other ways. By 2023, chatbots are everywhere, including social media, phone routing, and information services, where they were already becoming prevalent in our own day. Virtual personal assistants, or VPAs, have become ubiquitous, showing up in everything from kiosks to smart appliances to the increasingly autonomous vehicles that are becoming more common on the roadways with each passing year. VCAs, or virtual customer assistants, are the first line of help on the phone, the internet, and throughout our real-world urbanscape. Of course, it seems like almost everyone has their own favorite assistant on their phone, laptop, and voice-enabled home devices.

But beyond VPAs and VCAs, speech is driving the development of a new generation of voice-secured wearables, from health aids to smart jackets to noncritical medical devices. During the 2020s and much of the following decade, in any situation where control and output can be managed without the specificity of a mouse or a keystroke, voice will be the default interface.

It is a little as if we have traded roles, we and the machines. During the early days of computing, when computers didn't have any processing power or memory to spare, we picked up the slack in our new relationship by going the extra mile to learn and use languages and methods of communication that were extremely foreign to us. But as computers ascended the exponential curve of progress, more and more spare processing cycles could be used to make communications in our developing relationship smoother and ultimately more "human."

It is worth noting that the appearance of a new interface does not necessarily result in abandoning all previous ones. Just as the development of the graphical user interfaces didn't render the command line entirely obsolete, GUIs won't be abandoned just because of the arrival of natural user interfaces, including speech. Each interface has its time, place, and appropriate uses, and as long as this is the case, they will probably remain channels for human-machine interaction.

Back in the present, there are already millions of bots on the internet. By some estimates, as much as an astounding 52 percent of all traffic on the internet today is generated by software agents collectively known as bots. Good and bad, these range from crawlers that mine web pages to bots that retrieve information about products and services. The chatbots that engage with users using natural language are one rapidly growing subset of these agents. In 2016, Facebook Messenger launched its chatbot platform. By that September, there were 33,000 monthly active Messenger bots; half a year later, that number had tripled to 100,000, and a year later, in mid-2018, it had tripled again to 300,000.[8] By 2020, it is estimated that 80 percent of all businesses will be using chatbots, with a savings estimate of $8 billion by 2022.[9] The global market for chatbots has been forecast to grow by a compound annual growth rate of 24.3 percent to $1.23 billion by 2025.[10] By dealing with everything from personal scheduling, reservation booking, and maintaining task lists to enterprise processes of automating supply ordering, client invoicing, payment of vendors, and human resources tasks, chatbots and other software agents will come to automate an enormous proportion of our daily tasks. This will no doubt lead to substantial job losses, especially in help desk and customer service roles. But the convenience, efficiency, and cost savings from these bots will soon make them the consumer's channel of choice in many situations. Such are the market forces driving the accelerated development of this form of artificial intelligence.

There are hundreds of chatbot development platforms, all with their own approaches, weaknesses, and strengths. But for the most part they fall into two camps, represented by the rules-based scripting used by Pandorabots and neural network-driven machine learning systems such as Replika. The rules-based platforms are generally considered stronger for structured processes, such as technical assistance and customer support, while the machine learning approach is designed to evolve and self-modify according to user input.

It seems nearly impossible to discuss the future of chatbots and conversational AI without bringing up Spike Jonze's 2013 movie *Her*, in which an AI named Samantha becomes romantically involved with the movie's protagonist, Theodore. The viewer is quickly left with the impression that Samantha is conscious and self-aware, in large part because the AI's responses seem so natural. But we have no way of truly knowing whether Samantha is actually conscious or is simply doing a great job of faking it. Regardless of whether Samantha is or isn't conscious, we see by the end of the film that whatever this intelligence is, it is extremely different from ourselves. While it has been designed to connect with its user in the most natural of ways, we don't actually know that it is experiencing any of the interactions with Theodore as we expect another human would, no matter how well it emulates emotion and empathy. Because of the differences that have to exist between human and machine intelligence, it is fairly safe to assume that Samantha is simulating the emotional qualities that make Theodore fall in love with her.

The illusion of awareness will continue to be an issue with chatbots and virtual personal assistants for some time to come. Though they lack self-awareness in any human sense of the word, this doesn't mean they won't get very good at simulating it, which could lead to some people bonding with such devices in ways that may not be deemed emotionally healthy. Such a connection could leave the user vulnerable to exploitation, particularly in light of the fact that these interfaces will ultimately be commercial devices and services created for the purpose of growing profit margins. What will we do to protect ourselves from such manipulation? This is not so much an issue of potentially being taken advantage of by an AI than of a person using the AI for their own selfish gain, at least in the near term.

Advances in chatbot technology raise many concerns with respect to personal, financial, and psychological safety. Chatbots will surely be leveraged to manipulate people to greater and greater degrees over time. The potential for drawing someone into physical danger will be one type of risk. Another will be all kinds of financial fraud, from advance-fee and unclaimed property swindles to romance scams.

For example, in recent years the grandparent scam has become prevalent. This involves a youthful-sounding caller who phones an elderly person pretending to be their grandchild, who has been arrested and thrown in jail. They say they're afraid to call their parents but need to have bail money wired to them. If we skip ahead a few years, it's easy to imagine someone using chatbot technology along with "deepfake" voice generation to imitate the grandchild's voice, and perhaps even image, nearly perfectly. Combine this with information gleaned from the web and Big Data, then automate the process to rapidly contact millions of prospective victims. Only a small percentage would need to fall for the ploy for the scammer to realize a huge payday.

Then there are the psychological implications. For those who are impressionable, unbalanced, or simply vulnerable and lonely, the potential problems that could arise from long-term use might be enormous. A child raised with one of these devices may find himself unable to distinguish (or care about) the difference between dialogues with people and those with bots, unable to recognize the importance of consciousness in their conversations. Will later generations consider this a problem? Or will we find ways to adapt and deal with these nonhuman companions?

The problem of simulated intelligence is clearly illustrated by the thought experiment by philosopher John Searle known as the Chinese Room. Searle's hypothetical premise is that a program that receives Chinese words as its input and outputs appropriate Chinese words in response need have no understanding of the meaning of either. The process could easily be done manually by someone sitting in the room, having no knowledge of Chinese and only a book that tells him which characters to output in response to any given input. The entire transaction takes place in the absence of awareness, understanding, or consciousness.

However, the Chinese Room doesn't necessarily mean that the mental states we associate with consciousness will always be beyond the capabilities of AI. While there could be reasons artificial intelligence will remain devoid of the capacity for self-awareness or meta-cognition (thinking about its own thoughts), there are probably as many reasons to think that attaining self-awareness is inevitable. Most likely, machines will first achieve something comparable yet quite different from human consciousness, beginning as they do from nonbiological origins.

Regardless, the ongoing development of conversational technologies will in time result in virtual assistants so capable and personalized that we may think of them as sapient, even when we know they are not. We will come to wonder how we ever managed without them. Children will one day be raised with these devices as invisible friends that guide, teach, and protect them in a rapidly changing world. While comparable perhaps to the TVs and videos of recent decades, the intelligently responsive and interactive nature of AIs will make this a very different relationship. AIs will be upgraded multiple times throughout the child's life, growing up alongside the children in a manner of speaking. Children and AIs may become best friends and confidantes, in spite of, or perhaps because of, the differences between us. As the AI upgrades, it will presumably become more capable at recognizing the user's mood and modeling their state of mind. In time, AIs may themselves become self-aware or conscious in one or more senses of the word. But before this happens, there will be many challenges to overcome, as we will see in the next chapter.

CHAPTER 8

THE NEXT MILESTONES

"We shall say that a program has common sense if it automatically deduces for itself a sufficient wide class of immediate consequences of anything it is told and what it already knows. . . . Our ultimate objective is to make programs that learn from their experience as effectively as humans do."

—John McCarthy, computer scientist, author of "Programs with Common Sense," 1958

———

"Liam, can you find the doggy?"

The young woman, Chelsea, gestures at the menagerie of plastic animals positioned on the floor between herself and the toddler. Liam's gaze moves over the dozen or so toy animals before he points at one of them. "That's one! That's the doggy!" He snatches the German Shepherd in his tiny hand and pulls it to his chest.

"That's so good," Chelsea coos. "That's the doggy." She pauses a moment before asking, "What about the giraffe? Can you find the giraffe?"

The toddler spots the giraffe before she can finish the question. "Giraffe," he squeals, snatching it up and adding it to his collection. Liam beams.

"What about the lion? Where's the lion?"

Liam looks from one figure to another to another. His brow furrows as he starts to point at a house cat, thinks better of it, then scowls. "Lion's not here. It's gone."

"That's right, sweetie."

"Where'd it go?"

Chelsea extends her hand, revealing the lion that rests in her palm. "Peekaboo! There it is!"

Liam's face lights up with a great big smile as he grabs the lion from her. "Mine!"

The two laugh, continuing their play for another hour. Liam is as fascinated as he is tireless, but eventually Chelsea has to bring their playtime to a close.

The year is 2034, and Chelsea and her husband are part of a growing number of childless couples who have agreed to participate in Adopt-a-Bot, a government-funded AI training program. Liam is a lifelong learning robot.

Cute and a little cartoonish-looking, Liam and his "siblings" are a new generation of AI, designed to learn and develop much as children do. Entrusted to the care of parents around the world, these artificial intelligences incrementally acquire knowledge through their sensors and experiences. In doing so, they assemble views of and understanding about their worlds, in a process very similar to how young children learn.

Many challenges will have to be overcome in order to bring AI to this advanced stage. It certainly isn't yet possible, despite all of the amazing advances we have made in these first two decades of the twenty-first century. Several more milestones will need to be reached in order to move beyond the narrow AI of today and realize the general AI of tomorrow.

Narrow AI refers to artificial intelligence programs designed to perform a single task, such as playing chess, recognizing an image, or performing medical diagnostics. Most of the amazing recent achievements in deep learning use neural networks, such as Google's DeepMind program AlphaGo, which beat the reigning world Go champion in 2016, or the self-driving cars that are incrementally making their way onto our roads.

But human beings, and even a great many animals, have a much broader intelligence than any of these devices, which allows us to navigate the unending and multifarious challenges we face in the world each and every day. Without this capability, we would all have perished long ago, victims of what would have been very limited and inflexible minds.

This is why many AI researchers have begun to focus on developing ways to bring a range of broader intelligences to computers. Several are being explored concurrently by different labs and researchers around the world.

Many of the amazing achievements in AI during the past decade owe their success to deep learning neural networks, a model inspired by the way neurons are connected by dendrites and axons in our own brains. Neural networks are sometimes diagrammed using layers of circular nodes that represent artificial neurons, interconnected by lines.

Deep Learning Neural Network

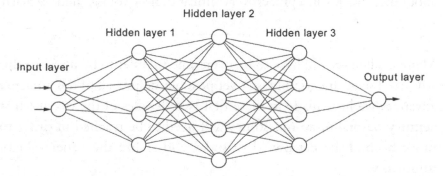

This is only a general representation, and there are many different types of neural network architectures that are used for different purposes in industries ranging from aerospace and manufacturing to medical uses and business.

Once established, a neural network is presented with training data, which is used to adjust the weights or values of the individual nodes and connections. Originally done as a supervised process, many of these adjustments are now made automatically using what is known as unsupervised training. Each hidden layer performs a role in detecting features in the data until the task is complete. Depending on the task, this may be followed by a classification stage that correlates the output with specific labels. As a result, such neural networks can range from two or three up to dozens of layers. Infrequently, the number may climb even further. Referred to as *connectionist AI*, neural networks are statistical in nature, with identification at each layer occurring with higher or lower degrees of probability.

One characteristic of many deep learning neural networks is their vast appetite for training data. These systems can learn to distinguish an image of a cat from that of a dog or to recognize speech, but this often requires huge data sets of tens of thousands, even millions of training examples in order to achieve accuracy. This data is used to incrementally refine the values and weightings of the nodes and connections that make up the layers of the network. Ideally, as the number of samples climbs, so too does the accuracy of the system. Because the deep learning training method is so dependent on this data, volume and quality are both critical.

Now compare this to how a child learns. Typically, once children reach a certain age and cognitive ability, it's simply a matter of presenting them with a handful of examples, and that is all their young minds need. Show a three-year-old a few pictures of kitties or puppies or rhinoceroses, and they'll know how to recognize that animal, in any light, from any angle, forever after. In the above scenario, this is one type of learning that Chelsea and the researchers are trying to accomplish with Liam.

Many milestones need to be met in order to build an AI that can begin to approach even some of the capabilities of a toddler. The ability to learn from a small handful of examples is what's known as "one-shot learning," and in 2020, we are only beginning to see this become a possibility for computers.

In artificial intelligence, one-shot learning is seen as a means of emulating or at least getting closer to the way humans learn using something called object classes. It is thought that by the age of six or seven, a typical child has acquired the ability to recognize somewhere between 10,000 and 30,000 object classes.[1] This allows them to readily distinguish "object class dog" from "object class cow," for instance (though, of course, people don't explicitly use such classifications). At the same time, this form of learning is predicated on prior knowledge, so that the ability to recognize "object class dog" and "object class cow" makes it possible to more readily learn to distinguish and identify "object class horse" from a limited set of examples. Not only would one-shot learning vastly improve the ability of AI to process, recognize, and learn from new and unique inputs, it would mean it had acquired the ability to learn from experience and extrapolate based on prior knowledge.

One method of one-shot learning being worked on for machines uses what are known as Bayesian frameworks that determine whether something is likely to fall into a particular object category or not. Proposed by Fei-Fei Li, Rob Fergus, and Pietro Perona in 2003[2], this approach gave a significant boost to one-shot learning research.

Another approach to achieving one-shot learning is the use of Siamese neural networks.[3] This involves two identical sister networks with identical initial weighting of their nodes and connections. The general strategy is to train the model to distinguish between a collection of pairs, some of which are the same while others are different. Each neural network is fed an input, perhaps one of two animals, two faces, or two objects. The systems are then self-trained to differentiate between the images. Afterward, this training is generalized to evaluate new categories based on what has been previously learned, making further learning possible based on deduction from prior knowledge.

Being comparative in nature, one-shot learning approaches lend themselves to building new knowledge from what is already known, a key feature in human learning and a desired capability for artificial intelligence.

Related to one-shot learning is zero-shot learning: the ability of systems to identify previously unseen object classes. The idea is that by using labeled training sets that classify different features of known and unknown objects, it should be possible to identify an unfamiliar one. For instance, using this method it becomes possible to recognize a male lion without ever having seen one before, because you know it is supposed to look like a large cat that has a mane of fur around its head. Both the zero-shot and one-shot learning methods build on prior knowledge, a key aspect of animal and especially human learning, but which has been challenging to fully implement in machines so far.

Another milestone that must be overcome is that of semantic understanding. When you or I say a word, we know that it may have different values and associations dependent on the situation as well as on other words that occur in proximity to it. For instance, the word *rose* may be a flower, a color, a person's name, or even the past tense of the verb "rise." Its meaning, value, and other qualities, as well as its many associations and connotations, are greatly determined by the language that surrounds it. On a number of levels, the context is central to the word's interpretation.

Words carry explicit and implicit meanings that we grasp fairly readily due to our experience with them, including our acquisition of them. But for the most part, early AI has been unable to deal with this, largely because it is unable to experience the world in a phenomenological sense. Instead, AI must extract meaning from words using methods in keeping with its strengths. Building meanings and associations with statistically dependent methods that use word tags and vectors allows AI to deliver an approximation of what humans readily achieve via our natural mental processes. The ability to truly sense the world gives us the advantage in encoding experience into language.

Semantic analysis is another, more statistically based method that is better suited to AI. This technique builds a hierarchy of related terms, meanings, associations, and conditions, which makes it easier for machines to infer the most likely meaning of a word relative to its context. The goal is contextual understanding. The ability to consider a wide array of relevant information is important in all communication, but especially critical where people are concerned. When two friends engage in a conversation, each phrase may consider the previous sentence, information from fifteen minutes earlier, and what's known about each other from five years before. Without context, such exchanges would be vastly poorer and more shallow.

In contrast, the chatbots and conversational interfaces of today consider only the previous sentence, perhaps two if we are lucky. Getting AI to fluidly consider such a broad set of unstructured data will likely need much more processing power and memory than is currently available. Recurrent neural networks may be better suited to working with data that represents changes over a span of time, but they need relatively structured data to work with, while real-world data is largely unstructured. That makes such a technique difficult to implement.

Perhaps a far greater challenge is the extent of the information that needs to be considered. Currently, deep learning AI algorithms have gotten amazingly good at specific tasks that fall within a limited domain. However, lack of contextual awareness limits just how useful and personalized these algorithms can be. For instance, a smartphone might automatically tell you the traffic conditions based on the fact you drive from work to home every day around five o'clock. But this is a far cry from an AI that responds usefully to a question like, "Where should I go skiing this year?" because it knows what snow conditions you like, what your typical budget is, and which friends might be available to join you. A lot of disparate information would inform this answer, and that is culled from a far larger pool of potential information. Such an extensive context would be necessary for a virtual personal assistant to perform roles equivalent to those done by a friend, colleague, or assistant.

Achieving more capable AI and eventually artificial general intelligence will probably require what is known as sequential learning. For many neural networks, learning a single task or process, such as playing a video game, is mostly a matter of training. Unfortunately, once that game has been learned, if the neural network is then trained on another game, even a related one, it can't carry over the knowledge or strategies it has previously accumulated. It effectively starts from a blank slate, experiencing what is known in AI research as "catastrophic forgetting." This is a far cry from the way people learn and build knowledge by incrementally accreting information, concepts, and relevant experience across a range of diverse tasks, and by using various tools such as analogy and inference to extend that knowledge. Recently, work has been done on a form of neural network that can learn several games in succession without entirely forgetting what it has previously learned. The system's ability to "remember" is due to the fact that it progressively builds and stores accumulated knowledge as it goes.

In 2017, Google's DeepMind team published a paper in which they showed how "continual learning" could be achieved based on "synaptic consolidation."[4] Inspired by how our own brains learn and retain knowledge, synaptic consolidation allows the network to learn a game, then to move on to a new game while transferring what it has learned. This results in more rapid training, because some of the prior knowledge doesn't have to be relearned. The way a neural network is structured means that this newer method doesn't play either game as optimally as it would if trained exclusively on a single game, but this is a step in the right direction.

AI faces a good many challenges before it can begin to compete with humans in terms of so-called general intelligence. While AI may be able to parse through millions of records in seconds or rapidly master a game that might take someone a lifetime to learn, it still can't shift from driving a car to negotiating an unknown building layout to running a board meeting all in a matter of a few minutes or seconds. Before it will be able to get close to that degree of versatility,

significant advances will need to be made in these many areas of contextual reasoning. That challenge is part of DARPA's motivation for
getting involved.

DARPA, the Defense Advanced Research and Projects Agency, is
responsible for developing emerging technologies for use by the US
military. While the involvement of the Defense Department and the
military in AI research and development may run counter to some
people's sensibilities, it's important to remember that the government's clout in funding can lead to new capabilities in the interest of
national defense, which over time often yield a significant dividend
for the civilian world, as has happened in the past. For example, a
communication system originally designed for connecting the military with other government departments as well as research universi-
ties was built by DARPA in 1969. ARPANET, as it was known, was
eventually turned over to civilian control, leading to the internet and
later the World Wide Web. Similarly, in the early 2000s, in response to
the difficulty of operating vehicles during the Iraq and Afghanistan
wars, DARPA initiated a series of grand challenges to bring together
teams of robotic engineers to develop autonomous vehicles, otherwise known as self-driving cars. DARPA's involvement accelerated
the development of these vehicles, probably advancing the state of the
technology by several years.

The success of these and other such programs led DARPA to
announce its AI Next Campaign in September 2018.[5] Designed to
address many of the hard problems in artificial intelligence, AI Next
has the goal of advancing the state of AI technology with a two-
billion-dollar investment over the next five years. The dozens of
separate AI programs seeking to address AI's challenges include projects such as "Accelerated Molecular Discovery," "Causal Exploration,"
"Explainable Artificial Intelligence," "Intelligent Neural Interfaces,"
"Lifelong Learning Machines," "Machine Common Sense," "Software-
defined Hardware," and "World Modelers."

While the initiative focused on most of the new AI capabilities discussed here, others were also deemed important to the future of AI and,

of course, to the military. These capabilities comprise what is being called the third wave of AI. The first wave corresponded to the early days of artificial intelligence, based on handcrafted symbolic logic and rules-based systems, a stage also known as Good Old-Fashioned AI, or GOFAI. The second wave encompasses neural networks that are connectionist in nature and much more statistically based. With the third wave, DARPA seeks to achieve contextual adaptation, a level of reasoning and abstract cognitive capabilities that would put AI much more on a par with the human intellect. If all goes according to plan, where the first wave *describes,* and the second wave *categorizes,* the third wave will allow AI to *explain.*

To achieve this, DARPA seeks to build contextual understanding and develop AI's ability to use contextual and deductive reasoning, abstract thinking, and *explainability.* While reasoning and abstract thinking are fairly familiar since they are so much a part of our own intelligence, explainability is something quite different. One ongoing issue in artificial intelligence is that neural networks are considered to be "black boxes" because of our inability to understand how they achieve the results that they do. Their opacity is in large part due to their statistical nature, as well as their being built around "hidden" layers of networked nodes. This is not to say that we don't know where the layers are; rather, what is unclear is how and why the network balances the layers as it does in order to achieve the desired results. This opacity persists whether the training is supervised or unsupervised. The processing behind the resultant systems is so complex that it is effectively impossible to follow step-by-step in a way that provides a logical explanation for the output.

This lack of explainability leads to all manner of potential issues, in areas from automated control of critical infrastructure, to accountability, which is essential for military operations, to unanticipated biases, be they racial, gender, socioeconomic, or otherwise. Being able to design a system that is able not only to perform a task but to explain how and why it did it will increase our confidence in our systems and, beyond that, effectively make them, the developers, and their users more accountable. Of course, DARPA's goal is far easier to state than it will be to implement. According to DARPA, they are working toward "a future in which machines are more than just tools that execute human-programmed rules or generalize from human-curated data sets. Rather, the machines it envisions will function more as colleagues than as tools. Toward this end, DARPA research and development in human-machine symbiosis sets a goal to partner with machines."[6]

Perhaps one of the greatest challenges will be in developing commonsense reasoning for these systems. Common sense is so innate to us, so based in our existence in the everyday world, that what seems like it should be no challenge at all is definitely one of the hardest challenges. As a simple example, we know that raw eggs have a shell that is fragile, something that is easy enough for a computer to look up in Wikipedia. We also intuitively understand the nature of gravity, which a computer can also easily look up and even simulate with a physics engine. Now if I tell you to hold an egg six feet off of the ground and release it, you have a pretty good idea what will happen. But, here in the early twenty-first century, computers simply don't have a clue.

DARPA is hardly the only group trying to resolve this. For more than thirty-five years, Cycorp's Doug Lenat has been working on Cyc, a project that has sought to instill computers with common sense.[7] While many contemporary approaches seek to automate the emulation of human thought processes, Cyc seeks to achieve this by producing an enormous number of handcrafted statements. (It's estimated that the Cyc project has assembled twenty-five million commonsense rules since it began in 1984.) Built around an ontology—a comprehensive data structure—along with a knowledge base and inference engine, the project has been trying to construct the rules of thumb needed to represent the common sense we routinely assemble in the course of living our lives.

Another organization working on this problem is the Allen Institute for Artificial Intelligence (AI2), established by Microsoft cofounder

Paul Allen and headed by computer scientist and entrepreneur Oren Etzioni. Among the institute's many programs, their Mosaic project is working to build commonsense capabilities into AI. To this end, the team began by developing ways to standardize the measurement of this very human property.

"Common sense is kind of like the dark matter of AI," Etzioni observes. "It's omnipresent, but it's kind of ineffable. Mosaic's starting point was to make it measurable, and now that we have some sort of a handle on that, we're looking at building repositories of common-sense knowledge. These are very different from the old Cyc project, which was manual and highly declarative. We're actually leveraging deep learning there."

Led by University of Washington computer science professor Yejin Choi, Mosaic is drawing on the world's best resource of common sense: people. In one project, they are using Amazon's Mechanical Turk[8] system to pay people to generate commonsense statements, building what Choi says will one day be the underlying rules that are needed. This will require a significant use of crowdsourcing, and Choi estimates she'll need a million human-generated statements to train her AI. Once in place, these statements of common sense could be used in an engine able to learn to predict and infer based upon the information.

Another AI2 program designed to learn common sense from interacting with people is Iconary, a game inspired by Pictionary. The game is played online by the computer and a person.[9] One draws, and the other guesses the phrase that is being represented. Not only does this take considerable NLP and vision recognition processing, but it also requires common sense. Both successes and failures incrementally teach the AI about different aspects of the world in the course of playing the game.

Google's DeepMind is another group working on these third-wave problems, building on the deep learning tools DeepMind has already had so much success with. Working with what's come to be known as neural architecture search,[10] or NAS, this system starts off by defining building blocks that could potentially be used for the neural network. The system puts these building blocks together to form an end-to-end architecture, then trains and tests it. Doing this multiple times, it eventually develops a relatively optimum configuration. Unfortunately, such an approach is very time- and resource-intensive, which has led to other approaches that attempt to explore the potential search space in a much smarter way, such as progressive neural architecture search[11] (PNAS) and efficient neural architecture search[12] (ENAS).

Building from there, Google and others have developed different versions of auto-machine learning, or AutoML. AutoML is designed to automate many of the processes of machine learning that ordinarily would require significant expertise to set up and tune. This sort of simplification has happened repeatedly in the evolution of computing and programming, where techniques are used to abstract many of the more challenging details of a process, making it possible for users with far more basic skills to create artificial neural networks. Beyond this, it should one day be possible for computers to generate on the fly the exact types of ANN modules that are needed for a given task.

Much of common sense is based on our experience of the world. Some is acquired through socialization, and the rest is the result of the realities of the physical world. Drop an egg and it breaks. Stub your toe and it hurts. This is partly why some computer scientists believe the answer to advanced AI is to raise it as you would a child. Let the AI incrementally build a wealth of knowledge based on what it learns from interacting with the world.

Linked to what we call common sense is the ability to comprehend causality, something AIs still don't do very well. The nature of causal relationships is critical to understanding the world, its actions and processes, and their repercussions. Understanding the implications of dropping an egg is a prime example. Or what happens if you are caught in a rainstorm without an umbrella. Or the consequences of driving a car over a cliff. Grasping the cause-and-effect relationship of these events is practically second nature to us. Artificial intelligence must be able to do this as well.

One approach that could help AIs learn causality is lifelong learn-ing machines (termed L2M at DARPA),[13] an approach that attempts to mimic human learning. Program manager Hava Siegelmann describes L2M as being structured around the four pillars of lifelong learning: continual learning, internal exploration, context-modulated computation, and new behaviors based on the accumulated knowl-edge. According to Siegelmann, "With the L2M program, we are not looking for incremental improvements in state-of-the-art AI and neu-ral networks but rather paradigm-changing approaches to machine learning that will enable systems to continuously improve based on experience."

Among other things, lifelong machine learning advocates want to move away from the enormous training data sets currently used by neural networks. Instead, methods that make much greater use of learned categories and knowledge transfer, such as one-shot learning (or even zero-shot learning), will probably play a significant role in implementing this method.

Some psychologists maintain we begin life hard-coded with certain key instincts. For instance, Harvard cognitive psychologist Elizabeth Spelke has explored the idea that children are born with a number of core knowledge systems[14] and that it is from this initial state that our abil-ity to acquire knowledge across our lifetime is bootstrapped. Building on Spelke's insights, cognitive scientist, author, and entrepreneur Gary Marcus has discussed the foundational distinction between "prewir-ing" and "rewiring," arguing that the biological evidence for both is significant. Marcus suggests there are probably a minimum of ten underlying human instincts or "innate traits"[15] that need to be seeded in AI for it to be capable of learning in a way that is more on par with the intellectual development of a child.

In contrast to this perspective, deep learning legend and head of Facebook AI Research (FAIR) Yann LeCun has indicated that repli-cating innate human traits would be a suboptimal approach and that the answer to achieving humanlike intelligence lies almost entirely in deep learning. This view is grounded in the assumption that all forms of intelligence have connectionist origins, as represented by the neural connections of the brain.

Each camp makes a strong argument for the best approach in getting AI to learn more like people do. Marcus states that convolutional neural networks, one of LeCun's greatest achievements, already incorporate one of the core instincts described in his list of ten needed innate traits, "translational invariance." In artificial neural networks, translational invariance allows an image to be detected and identified regardless of where it appears in the field of vision or its orientation. LeCun has countered the idea of preprogramming this and other innate traits, saying that such translational invariance could arise naturally, given more sophisticated deep learning methods. Is it possible each could be right in their own way? Could the formation of different biological neural structures and processes have been optimized through evolution to be "primed" to fulfill the role of these core knowledge systems?

Many major challenges in AI are the focus of research labs around the world, but there can be no doubt it is the tech giants that have the upper hand. Those companies have the resources to hire the best AI talent as well as access to the enormous pools of data their companies generate. Given the success of deep learning, Google, Facebook, and Amazon are likely to remain the leaders in this field for a long time to come.

So too with China's tech giants Baidu, Alibaba, and Tencent. Collectively known as BAT, in 2017, these three companies had a combined value of over $1 trillion.[16] Though their valuations have since dropped significantly, these companies still remain huge players in the field of AI and, more specifically deep learning.

For decades, China has been criticized by the West for what many consider its restrictive and authoritarian approach to the internet. Filtering major sites and services from its citizens, keeping them behind what has been ironically dubbed the great firewall of China, the government's policy has the potential to put the country at a disadvantage in knowledge sharing and innovation. But with the development of deep learning neural networks, it has become evident that China's situation is anything but a disadvantage. Deep learning is extremely data intensive, and with a population more than quadruple that of the United States, these Chinese giants have continual access to as much potential training data as they could want. Besides having full access to more than 1.4 billion citizens, 1.1 billion of whom are on smartphones, these Chinese companies have a rapidly growing user base outside the country as well. Tencent, which operates the messaging, social media, and mobile payment app WeChat, has nearly 1 billion monthly users. E-commerce giant Alibaba processed over $8 trillion in online payments in 2017. Meanwhile, Baidu, who fulfills three-quarters of all search engine queries in China, performs more than half a billion searches daily. While less than one-tenth of Google's daily six billion searches, this still represents enormous opportunity, as Baidu's annual revenue of over $15 billion attests.

Besides building AI to support their core businesses, these companies are rapidly expanding into countless markets made possible by the ongoing developments in artificial intelligence. As well as developing superior machine translation, they are putting autonomous vehicles on the road, neural chips in robots, and virtual assistants into homes. Competing with Amazon's Alexa and Google Assistant, Baidu's DuerOS-enabled voice assistants reached 100 million units in 2018, having doubled its numbers in only six months.[17] Additionally, all three giants have made major commitments to facial recognition, with Baidu famously foregoing all employee badges for a facial recognition system they consider far more reliable.

Do these numbers point to a future in which China is the dominant leader in AI, particularly deep learning? Quite possibly, but there remain many unknowns as the different anticipated breakthroughs in areas such as contextual reasoning and one-shot learning are made. AI-based translation is already more accurate than human translators in some applications. This is in large part because advanced systems have moved from statistical phrase-based machine translation (SMT) to neural machine translation (NMT). In SMT, a statistical approach is used to determine the translation of words and phrases. Using recurrent and convolutional neural networks, NMT does a far better job of accurately conserving meaning and using correct grammar.

There is another reason to think China might have an advantage in the AI race. Contextual understanding is important in achieving truly superior translation, which, as we discussed earlier, is still waiting to be realized. While Mandarin and Cantonese are notoriously challenging for non-native speakers to learn, their grammatical complexity is less daunting than English, making contextual understanding potentially easier as AI develops. This is important. Learning words and language in context is a universal phenomenon for people but far more challenging for AI, at least until some of the anticipated advances of the 2020s have been made. In the meantime, the individual components of Chinese characters provide categorical information about their meaning,[18] potentially rendering them more easily understandable than their English counterparts.

In many respects, the future of several of these AI challenges are interlinked. Advances in contextual reasoning, one-shot learning, semantic understanding, explainability, and common sense will probably occur in these individual domains as well as interdependently, linking up with advances in each area. Will the idiosyncrasies of the Chinese language be what provides the advantage to China in bootstrapping the next wave of AI advances?

However we look at it, sharing our future with artificial intelligence will result in a very different, often strange world. The existence of systems that can read, interpret, and anticipate us will be empowering, making many aspects of our world that serve us seem almost like magic. Environments everywhere will listen to our needs—in those cases when they haven't already anticipated them. Vehicles will adjust their driving style if they're making us too nervous. Our clothes will let us know when we last wore them and who saw us wearing them, as well as offering other wardrobe advice.

But not everything in this new world of wonder will be for the better. Soon we will be seeing empathy bots that use sob stories to convince us that they are actually human, in an effort to scam people out of their money, security information, or both. Every election cycle will be vulnerable to the next escalation of weaponized social engineering. AIs capable of replicating people's voices, biometrics, and personal information will make regular attempts on people's life savings.

It will definitely be a matter of taking the good with the bad, which is why we need to plan ahead as we develop these increasingly intelligent technologies. Anticipating the vulnerabilities of these systems as well as understanding our own values, priorities, and vulnerabilities will be essential if we are to build a future that we actually want to live in. Change is happening too rapidly to depend on the ad hoc approaches of yesteryear. We can no longer afford to have a mindset of "build it and let later generations sort out the mess." Instead, it will be necessary to develop a methodological approach for better assessing the environmental, societal, economic, and ethical transformations all new technologies and their applications may lead to, the object being to apply the insights gained from the process as early as possible.

So, you may wonder: if that is the case, why develop these technologies in the first place? Why not ban them entirely, or at the very least put a temporary moratorium on them until we can better prepare for their impacts? Unfortunately, such an approach will not long succeed even in a highly controlled, authoritarian state, much less in a free and open one. We live in a global society in which economies and knowledge are highly interconnected. Once knowledge has reached a certain level and the necessary infrastructure has been developed, if a new technology makes economic sense, it will be built. There is no stopping this process, not unless all parties decide it is in all of their interests to stop. To take half measures in which only some nations ban a development will result in R&D being forced underground in some places and pursued openly elsewhere. If relegated to the shadows, the outlawed technology will be unmonitored and unregulated. Furthermore, benefits gained by those who do not adhere to the ban will result in a competitive disadvantage to those who do. Depending on the technology, and especially in the case of artificial intelligence, this could have a tremendous impact on a nation's standing in the world.

It is worth mentioning that throughout the remainder of this book, technologies and trends will be extrapolated in order to anticipate how our future world might develop. In discussing them, I will point out both positive and negative aspects, but generally speaking, it is not the purpose of this book to advocate for or against any of these possible developments. We are unlikely to halt any technology from being realized once its time arrives. What we must learn to do, both as individuals and as a society, is prepare for the coming changes and adapt our systems and behaviors in order to continuously, iteratively shape the kind of world we want to live in.

Our values and vulnerabilities need to be carefully taken into consideration as we move forward in developing artificial intelligence. For decades, the useful advances in this field have been almost exclusively in weak AI, systems that perform a limited range of tasks in a highly restricted domain. Image recognition, fraud detection, and recommendation systems are all examples of weak AI. But with the capabilities that will be developed over the next decade, we are rapidly entering an era in which artificial general intelligence (AGI) will become not only possible but potentially prevalent. Unlike the narrowly focused AIs we have developed to date, AGI systems could eventually be built that are capable in a far broader sense. Not only would these AGIs be able to solve problems much more as people do, but their potential for directed self-improvement could make them a significant threat. With the development of AGI, we are likely to face a host of new dangers and challenges, and there is little hope this can be forestalled. The major powers of the world, and particularly their militaries, are convinced that AGI is their future. Or, if they do not believe it, they at least believe that their adversaries do, leaving them with little choice but to pursue it themselves. As Russia's leader Vladimir Putin observed about AI in 2017, whoever "becomes the leader in this sphere will be the ruler of the world."[19]

As existential threats go, AGI has the potential to be as great as nuclear weapons, perhaps even worse. But quite a few other milestones must be met before this happens, as we will continue exploring in the pages ahead.

CHAPTER 9

GETTING EMOTIONAL

"Rather than being a luxury, emotions are a very intelligent way of driving an organism toward certain outcomes."

—Antonio Damasio, neuroscientist

———

The self-driving taxi sped down the freeway rapidly and efficiently, surrounded by a multitude of similarly rapid and efficient autonomous vehicles. All of them moved in unison, with less than a foot of clearance on each side. A solitary passenger sat in the taxi's backseat. Sensors estimated the man was approximately forty-five years of age and of height and weight within typical adult norms. He also appeared to be making a considerable effort to remain calm. The taxi noted this in its logs.

"Look out!" the man suddenly shouted.

Over the next few milliseconds, the taxi quickly triple-checked its proximity sensors as well as the visual and telemetry feeds of the traffic in all directions. All conditions were normal.

A soothing feminine voice emanated from the compartment speakers. "No threat is detected. Are you okay?"

The man continued to sit upright and rigid. "Yes. Yes, I'm fine."

The taxi's EMS—Emotion Monitoring System—contradicted the man's statement. It calculated with greater than 99 percent confidence that he was in fact highly agitated and scared.

"You're perfectly safe," the taxi stated. "I have an exceptional safety record. My entire company does."

The passenger remained silent.

"It's evident you're still upset," the vehicle continued. "Would you like me to pull over?"

Internal sensors indicated the man was now hyperventilating. The vehicle transmitted a shoulder request and the traffic around it opened up, allowing the taxi to move to the side of the road three lanes away, before decelerating to a stop.

"Take deep breaths," said the taxi. "You're safe. Everything's going to be all right."

The man took several slow, deep breaths, which seemed to help. The taxi knew that external appearances did not always perfectly correlate with a person's internal mental states.

"I'm sorry," the man said, forgetting he was talking to a machine. "I rarely use cars. They make me nervous."

"I understand," the taxi said. "Some people still aren't comfortable with autonomous vehicles. But we are so much safer than when people used to do the driving."

"I know the statistics. But that still doesn't make me feel any better."

The taxi registered the man's continuing trepidation and recognized this wasn't a worthwhile line of conversation. "I understand," it said, resorting to its default acknowledgment response. Momentarily, the passenger cabin was filled with soothing music: Pachelbel's Canon in D at low volume. The passenger released a long sigh and relaxed perceptibly. The taxi spoke again, incrementing the soothing cadence property of its voice generator.

"Why don't you close your eyes and focus on the music?" The man closed his eyes, taking a deep breath. After a minute, the taxi calmly asked, "Are you okay if we continue our journey? No need to speak. Just keep your eyes closed and focus on the music."

The man nodded almost imperceptibly and the taxi signaled for extra space, so it wouldn't need to accelerate too quickly. Pulling into the flow of traffic, man and machine continued on their way.

————————

As computer interfaces have developed over the decades, one particular trend has been more prevalent than any other: the progression toward ever more natural interfaces. From hardwired instructions, punch cards, and the command line interface to graphic user interfaces, touchscreens, and voice commands, our technologies are becoming easier and more natural to use, anywhere, anytime. Now we are seeing the beginnings of a new interface, one that interacts with some of the most basic aspects of human nature: our emotions.

From the dawn of humanity, emotion has been central to who and what we are. Long before we could speak or write, emotional expression was our means of communicating with each other and understanding what was going on in each other's minds. By reading the expressions on someone's face, the spring in their gait, the agitation in their gestures, we gained immense insight into what that person's intentions were. For all this time, interpreting these subtle signals has been an almost exclusively human capability. Until very recently.

In the mid-1990s, MIT Media Lab professor Rosalind Picard wrote a book that would launch an entirely new branch of computer science: *Affective Computing*.[1] The book and the field that followed from it involve a range of technologies that can read, interpret, and interact with human emotions. (The history, applications, and future implications of this field are explored in my 2017 book, *Heart of the Machine: Our Future in a World of Artificial Emotional Intelligence*.)

Following a decade of successes at the Media Lab, Picard and postdoctoral researcher Rana el Kaliouby spun off the emotion detection company Affectiva, based in Cambridge, Massachusetts. Initially focusing on wearable emotion detectors and facial expression recognition software, the company eventually decided to concentrate on the latter for the better part of a decade. Picard went on to found another Cambridge company, Empatica, in order to continue developing wearable emotion detection technology that is used to anticipate and monitor for epileptic seizures.

Throughout that period, numerous other companies entered the affective computing market, or, as the field also came to be known, emotion AI. Many of these companies focused on facial expression detection as Affectiva did. Some built their own technology from scratch, while others used the APIs (application programming interfaces) already developed by Affectiva and others in the field. Still other companies focused on different modalities such as detecting emotions in voice, posture, or through galvanic skin response. Because emotion is multimodal—meaning it can be expressed and read via different modalities, such as face, voice, and gesture—different technologies can approach the task using many different strategies.

Initially, affective computing was extremely challenging to develop and use due to the variability of people's expressions, vocal qualities, facial features, and much more. But with advances in artificial neural networks in late 2000s, tremendous strides were rapidly made. Once the commercialization of affective computing was established, market research companies such as Millward Brown[2] began to use the technology for consumer sentiment research and brand testing, seeking to detect subtle responses that might have previously gone undetected using only surveys and interviews.

Over the next decade, companies sought to apply this technology to everything from decision-making, fashion recommendation systems, and alertness detection in automobiles to customer relationship management, healthcare, and robotics. The more areas that were explored, the more it seemed to prove that nearly everything with which people interacted could benefit from incorporating emotional awareness.

Over time, companies building digital assistants also wanted to add emotional awareness to their devices, recognizing this could play a significant role in improving customer experience. This has led several AI giants to buy or build their own emotion AI services. In January 2016, Apple acquired affective computing company Emotient, presumably for its facial emotion detection technology. Microsoft has developed its Face API (formerly Emotion API) within Azure Cognitive Services, which incorporates a number of emotion detection and recognition routines. In November 2016, Facebook acquired Carnegie Mellon spinoff FacioMetrics, whose app recognizes seven different emotions, again with an eye to adding these features to its platform. Amazon's Rekognition technology analyzes live images and detects emotional attributes in people's faces. In 2018, Rekognition was the focus of considerable controversy, due to claims of racial and gender bias combined with Amazon's plans to provide the technology to law enforcement. In 2019, it was also revealed that Amazon was working on a project codenamed "Dylan,"[3] a health and wellness device that can be worn on the wrist and is able to detect the wearer's emotions.

Where is all this going? The day is fast approaching when many if not most of the devices we use will have some degree of emotional awareness. They'll alter help screens when they detect we are getting frustrated and offer perspective when we are feeling down. As Affectiva's Rana el Kaliouby has observed, "I think in five years, all our devices are going to have an emotion chip, and we won't remember what it was like when we couldn't just frown at our device and our device would say, 'Hmm, you didn't like that, did you?[4]'"

All of this will lead to a very different world, one which is far more responsive to our subtlest cues, even unconscious ones. Already affective computing is being used in market research, automobile safety, fashion, and autism detection. Versions of this technology have been tested and used in hospitals, education, and therapy. There are applications being explored in human resources, sales, and robotics and automation. Wherever there is a person to interact with, there is an opportunity for affective computing.

Of course, this also raises some very important questions. When and where will this technology be allowed to be used? If we are out in public and our facial expression is being read, is this an invasion of our privacy? If we determine that it is, how is this different from another person seeing our face and making a judgment about how we feel or our state of mind?

In an age when our personal data is considered a valuable asset and we are questioning corporations' rights to access and use it, what are the implications? After all, what data could be more personal than our emotions at any given moment, in response to any given situation or experience? In a 2018 survey,[5] global research and advisory firm Gartner confirmed this concern, finding that more than half of the more than 4,000 people they surveyed say they don't want their expressions to be analyzed by AI. This is particularly interesting, given that Gartner also anticipates that 10 percent of personal devices will have some emotion detection features by 2022, up from only 1 percent just four years earlier.[6] Not surprisingly, the survey also found that opinions differed based on the respondent's age, with millennials expressing fewer concerns about privacy if it meant having the additional conveniences promised by emotion detection technology.

Then there's the potential for manipulation. While we can frame this in terms of our being manipulated by these machines, what we really mean is manipulation by people using this technology, be they corporations, governments, or hackers. The fact that emotion is so central to who and what we are makes it a very powerful tool if it is used against us, even as emotion AI improves many other aspects of our lives and our world.

Yet affective computing technologies also hold the potential to bring many amazing capabilities and conveniences to our lives. As with so many technologies, how it is ultimately used will determine its benefits and risks. Identifying these early on and creating the needed safeguards is how we will ensure a more preferable future.

A technology like emotion AI could lead to a very different world during the next decade or two. As our technologies continue to shrink in size, they will disappear into our environment, becoming available anywhere, anytime. Our voices, faces, and no doubt a range

of other biometrics we haven't imagined yet will allow us to autho-
rize access to our files and personal data. We will have an array of
filters and layers of protection behind which we will decide what data
to reveal to which vendors and services. In spaces public and private,
different players will continuously seek to access whichever of our
emotional signals they are legally allowed to, all the while enticing us
with promises of conveniences, access, and more. It will be a trade-off,
a continuous game of give-and-take, just as it is today with our other
data, only more so.

Affective computing has only recently become feasible because of
advances in computing, particularly processing power and neural networks. We discussed neural networks earlier, but just how do they do
what they do? Artificial neural networks, or ANNs, are a group of
nodes and connections inspired by biological neurons and dendrites.
ANNs can be entirely hardware-based or they can be implemented
in software. These networks are typically structured in layers which
incrementally process the incoming data. In the case of recognizing
an image, the input layer feeds the data into the next layer, which
performs the initial level of pattern recognition, perhaps identifying
certain edges or lines. This is fed to the next layer, which finds some
more patterns, which are fed to the next layer, and so on. At the
end, after passing through several layers, the image is fully identified
and classified. This is a very simplified version of what happens, and
there are usually pre-processing stages that clean up the data, adjusting
aspect ratios, image scaling, and normalizing inputs. Post-processing
can also be performed for a range of other purposes.

Once they are configured, ANNs are typically trained by presenting them with large sets of data, for instance many different pictures
of cars. These and other images might be labeled "cars" and "not cars."
As the ANN is introduced to these images, it adjusts its "weights," the
values of the various nodes and connections in order to come to the
desired result.

Traditionally, the weights in an ANN had to be adjusted manually,
in what is known as supervised training, a slow and tedious process

that requires the attention of highly skilled AI scientists. But over time, methods such as backpropagation were developed that allowed many ANNs to be trained in a relatively unsupervised manner, effectively adjusting their weightings themselves.

Though the concept of ANNs had been researched for decades, it was the significant gains in computer processing power (thanks to Moore's law) that allowed this technique to become so powerful during the past decade. The immense processing needed to calculate several layers of nodes simply wasn't available in the 1980s and 1990s, and it is these layers that give neural networks their incredible abilities.

So how is this used in affective computing? In the case of facial expression recognition, images of people's faces are carefully selected and labeled according to the emotion being expressed. The data is then used to train the ANN, adjusting its many weights until the desired result is obtained.

One key feature of machine learning and neural networks is that they can be extremely data intensive. This was in fact one of the motivators for Picard and el Kaliouby to strike out on their own and launch Affectiva. In the confines of the MIT Media Lab, it was laborious work to generate even several dozen training images. But their work needed tens, even hundreds of thousands of images in order to truly be successful.[7] This is because these systems are essentially a form of statistically based learning, and so the more good training data they have, the better the system gets. Nor is this a linear relationship, since it tends to follow a path of declining returns. With a few dozen samples, you might get to 55 percent accuracy, which is only slightly better than chance, but 60 percent could require several hundred items. Reaching the 90 percent mark often requires many other tricks in the mix, but it especially needs a vast number of samples. By moving into the commercial space, Affectiva was able to build an immense training set, especially from the webcams of users who were willing to participate in the process.

We've been discussing ANNs in a somewhat generic sense, but in fact, there are many different varieties. Convolutional neural networks, or CNNs, are generally considered well suited for still images, but webcam and video feeds take place over time, and therefore moving images may benefit from using other types of ANNs, such as recurrent neural networks, or RNNs, which are better able to process temporal sequences. Emotion AI actually also benefits from this because of the ever-shifting nature of facial expressions and micro-expressions. None of us naturally expresses an emotion as a freeze-frame, but rather as a fluid set of movements that change from one fraction of a second to the next. Because of this, moving images can be far more accurate representations of expressed emotions.

Different techniques are used for different emotion modalities— voice, gesture, posture, etc.—but many build on the same tools. Vocal expression especially shifts and changes over time, making RNNs a better fit. One company that has been involved with voice emotion detection for many years is Tel Aviv–based Beyond Verbal. In their research, Beyond Verbal has uncovered many interesting commonalities between our voices, independent of language and culture. Recently, they engaged in a two-year study with the Mayo Clinic to show that vocal biomarkers can predict a number of chronic diseases, including coronary artery disease.[8]

Exciting as this new interface technology is, the potential relationship of affective computing to the future of intelligence may be the most intriguing. When we talk about human intelligence, most discussion is focused on our general abstract or fluid intelligence. Only relatively recently has the value of emotional intelligence even begun to be recognized. Of course, emotional intelligence has never been a consideration for AI before. But should it be?

The initial intended purpose of building emotional awareness into our computer systems was to improve the way they respond to the user. It was hoped that by recognizing when users were becoming frustrated or bored, a program could branch to an alternate action to reduce frustration or make itself more engaging. But as the technology developed, other applications became apparent. For instance, the ability of corporations to be able to glean new information from people in order to better target their marketing was a major attraction. In the era of Big Data, a few unique pieces of information can delineate an entire segment of the market.

In many respects, these commercial applications are much more the market drivers for this technology today, transforming these emotion interfaces from curious playthings into devices and services that companies are willing to pay huge sums for as the emotion economy develops.

Of course, a technology that probes the nature of our feelings is bound to raise many issues about personal privacy. Concerns about privacy are only going to increase with the growth of the Internet of Things, or IoT. Corporations have been talking about the promise of IoT for the past couple of decades. In recent years, this technology has begun to grow rapidly, with estimated worldwide spending on IoT expected to reach $1.2 trillion in 2022, according to market intelligence analysts at IDC.[9] As the number of sensors in our environment increases, we will find all kinds of new uses for them. Emotion AI will almost certainly be one of these.

Imagine you are shopping at a mall. The moment you enter, posted notices state that you recognize and accept that you are being monitored. As you approach a particular clothing store, its cameras identify the items you are wearing. Perhaps more importantly, it identifies your mood.

As you reach the store, a large video display projects images of clothing and products that have been algorithmically selected based on information the store has already gathered about you. As you stop and glance at the display, cameras rapidly assess your micro-expressions— unconscious muscle movements that correlate to specific emotions. Based on this feedback, the displayed products are altered, building your interest until you finally decide to enter the store, ideally to make a purchase.

In many respects, this is business as usual, since corporations have always sought to create enticements that promote their sales and feed their bottom line. However, the imagined system just described is much more granular. Instead of marketing to a large target audience with products and displays that appeal to a generic aggregate group, the combination of technologies in this scenario would make it possible to engage with shoppers on a highly personalized, one-on-one basis. Perhaps even more importantly, the ability to detect and respond to an individual's emotional responses on the fly creates a real-time feedback loop. It effectively results in an unspoken conversation between buyer and seller, one of which the buyer may not even be aware. For the AI, though, the conversation becomes an optimization loop for modifying human behavior.

Essentially, what is depicted here is a form of modifiable predictive analytics. In predictive analytics, computers mine data for patterns that allow them to make predictions about the future. In this case, the computer is drawing on a vast data set about fashion, demographics, and so forth. This is combined with information about what the shopper is wearing to make an assessment about what they like and what they are likely to buy. Then, as the interaction occurs, the program modifies its assessment based on any new emotional input it receives in order to make and exploit a more accurate prediction about the buyer's anticipated behavior.

Predictive analytics is an extremely powerful tool, as illustrated by research work performed in 2011 on a system called Far Out.[10] For the study, Adam Sadilek of the University of Rochester and John Krumm at Microsoft Research assembled a large data set from 703 subjects, made up of paid and unpaid volunteers who wore GPS trackers from many months. At the end of the tracking stage, the GPS units were removed. Nevertheless, upon plugging the gathered data into their algorithm, the researchers found they could predict where any of those participants would be at any hour of any day within a one-city-block radius with greater than 80 percent accuracy—up to a year and a half later.

Such predictive capability is incredibly powerful and is only going to become more accurate in a world of IoT sensors and neural networks that can find patterns where the human mind never could.

This is but one application of this powerful technology. How many ways will we find to use it, and what will be the long-term repercussions?

Now let's look at this from a different perspective. When computers examine people's behaviors, we talk about techniques such as predictive analytics and behavioral informatics (the use of data to understand and anticipate human behavior). But when people do the same thing to other people in order to model their mental states and motivations, we call this "theory of mind."

Theory of mind is something we human beings have experienced and used for a very long time, probably from well before our species even had the capacity for complex speech. The ability to anticipate what goes on inside another person's mind, whether friend or foe, has immense value in promoting our survival and competitiveness in the world. Over time, the more we learned to read and understand people's expressions and emotions and correlated this with what else we knew about them—their circumstances, values, motivations—the better we got at predicting what they might do next.

Most of us learn to do this from an early age, to greater and lesser degrees. It's such a basic part of our interactions with each other that most of the time we are not even aware that we are doing it.

But beyond this, theory of mind binds us to each other. Being able to model the mental states of other people promotes our understanding of them, our ability to see ourselves in their shoes, to empathize with them. With this ability, we established one of the key pillars of consciousness.

It's very challenging to draw direct connections, to say that one aspect of consciousness followed from another or even how it originated. What we do know, however, is that we are the only animal that exhibits such a high level of theory of mind and that it almost certainly has been instrumental in transforming us into the type of social animal we are today.

But as AI incrementally acquires the ability to perform pattern recognition, abstract thinking, reasoning, and common sense, is some form of theory of mind a possibility too? If so, could this be the precursor to something more experiential, something closer to self-awareness and other aspects of consciousness?

AI won't reason like us or identify patterns like us, so why should it experience anything like consciousness like us? It's safe to say it won't. At least not for a very long time, and by then, perhaps only as a curiosity. After all, the reason our form of consciousness seems so special to us is because that's all we know.

From an engineering standpoint, it is ultimately the result, not the method, that matters. Building an airplane or jet that can fly is of much greater importance to us than the ability to do this exactly the same way a bird does. In fact, by achieving flight differently, we are actually able to fly further, faster, and carrying far more weight than any avian ever could. In much the same way, by not tying the methods of computing too closely to their biological inspirations and analogues, our technology has been able to do far more—at least within certain activities—than we can ourselves.

But even if AI's theory of mind isn't like our own, it could still yield many dividends. Understanding the minds of others is a major component of empathy—cognitive empathy, to be more exact. At least for certain systems, the ability to empathize would be very useful and could go a long way toward protecting human users, both emotionally and physically. The ability to *simulate* empathy, on the other hand, could be a double-edged sword, given the option of using it for our benefit or against us. For instance, a therapy bot could be improved by simulating empathy. On the other hand, a robo-caller that used simulated empathy to prey on people's sympathies in order to trick and rob them could be a possibility as well.

Though it will be counterintuitive for us for some time, we must realize that the ability to simulate empathy and emotions does not mean an AI would actually experience any of these. While this may seem obvious and our rational minds may recognize it, our emotional selves will still want to respond accordingly, because in the past we have only ever had to interact with other humans in this way and because it is in our nature to respond to empathy reciprocally, as to a fellow human. But if a bot calls you on the phone and can carry on a conversation that is indistinguishable from another person, down to the voice, reasoning, and emotions, you will be inclined to respond accordingly, even if it does identify itself. Of course, if it doesn't identify itself, then you are clearly at a disadvantage.

That said, is there a future in which we would want AI to experience emotions? What would be the benefits? What issues and challenges must be overcome? Most importantly, what dangers, if any, would this bring?

Emotion, as I have explored elsewhere,[11] originates from our bodies. Our endocrine, sympathetic, parasympathetic, and enteric nervous systems all contribute to our basic emotions—whether we feel happy, sad, angry, scared, disgusted, etc. Our mind subsequently interprets and labels what we are feeling based on our prior experience and culture. Certain other emotions, such as shame and guilt, typically originate in our higher-order thought processes. While these appear to be more cognitively and less somatically based, it's unlikely the more enculturated emotions could ever have evolved on their own were we not the highly social creatures we are.

The physical origin of emotion points to the idea that without a body it might be impossible for AI to experience emotions as a person does. Needless to say, building a complex, multifaceted nervous system for an AI or robot would be daunting. But if we could do this or if we could find a way to create something analogous, what might be our motivations?

One answer might be found in the differences between how computers and people store memories. Traditional computer memory is highly detailed and conserved, so that for instance it is possible to know exactly what date, time, and location a picture was taken. This is very different from human memory, which retains information that is at once inferior to the computer's detail but also so much richer in other ways. While each form of memory has its merits, it would seem that the exponential growth of the amount of data generated by our technological world will make it increasingly challenging to incorporate so much structured, detailed data into anything modeled after human memory. Instead, a model that results in some greater or lesser degree of memory consolidation might be much more suited to certain (though not all) aspects of contextual intelligence.

Often our memories are closely linked to our emotional associations with that memory.[12] This is a result of our evolution, tying important moments to emotions during memory formation and retrieval, particularly strong emotions such as fear. The genes that made this possible would have likely been selected for because they promoted the retention of critical knowledge that might help us avoid or survive a similar situation or threat in the future. This emotional link to our memory and cognition also allows us to determine value throughout our experience of life and our environment—that is, value in the sense of knowing what is most important to us at any given moment in order to know where we should be putting our attention[13].

For instance, a computer might be the greatest chess player in the world, but if the room it is in catches on fire, it has no conception of this and will just keep playing chess until its circuits melt.[14] Our emotions are continually shifting relative to the shifting conditions of our environment. Because of this, our priorities can change in an instant, which in many cases is exactly what needs to happen.

Computer systems have no real means of knowing what to value in the moment or where to put their attention other than where they have been programmed to focus. It could be said that this is linked to their inability to experience the world or to have emotional responses to it. A robust AGI would need to be capable of taking in all kinds of unstructured information and adapting its responses to the conditions of the moment, much as we do. While human beings are far from perfect at this, we are so much better at it than a brittle computer program that fails at the first moderately unusual input. Though far easier to state than to implement, the capacity to experience the world would take AI a long way toward becoming more capable and resilient in its actions and responses.

Perhaps even more importantly, a degree of internalized emotional awareness could be used to align the reasoning and resultant actions of some of these systems with our own. This is known as the value alignment problem, and addressing it may be critical to the long-term success of both humanity and technology, as we'll explore in chapter 13.

Beyond this, we may one day have other reasons to create an AI that can actually experience a form of consciousness. Perhaps we would do so in our continuing efforts to unlock the secrets of human consciousness. Whatever our future motivations, the ability of technology to experience emotion may be critical not only for its continued advancement, but for humanity's as well.

CHAPTER 10

WHAT MAKES A MIND?

"Thinking is a physical process . . . The human brain is not exempt from evolution."
    —Steven Pinker, cognitive psychologist, linguist, author

———————

Rescaled to the size of a red blood cell, our miniaturized ship rapidly negotiates a vast branching cardiovascular system. The blood vessels fork and taper, eventually narrowing to a network of capillaries that restrict the circulation to ourselves and a single file of crimson corpuscles. The continuous surge and flow of the bloodstream is relentless and would no doubt leave us feeling seasick were it not for the ship's inertial dampers.

We've teleported into the brain of one of Earth's nearly eight billion citizens in order to find out what makes them self-aware, sapient, and conscious. It is a brain like most others. There's nothing particularly special or different about it beyond its uniqueness relative to every other brain on the planet. Anything we find here should be representative of almost anyone we might encounter. So we'll just take a quick look around, find the seat of intelligence, and go. Rest assured, no brains will be harmed in the writing of this chapter.

Approaching the brain stem situated at the base of the skull, there's nothing obvious to explain this person's intelligence or self-awareness, but that comes as no surprise. This region of the brain performs dozens of basic life support functions, much as it does for all vertebrates, being responsible for the autonomic nervous system, which includes breathing, heart rate, and blood pressure.

Moving toward the blood-brain barrier, a semi-permeable protective border, it quickly becomes evident that we're still too large by several orders of magnitude to pass across it into the rest of the brain. Shrinking our ship down by several thousand times, we slip through the barrier with ease.

We pass along the pons and the cerebellum as we continue our search. The pons relays signals between the forebrain and cerebellum and is involved in sleep, respiration, swallowing, and bladder control, among other things. The cerebellum is critical to motor control and coordination, as well as possibly to attention and language. Consisting of hundreds of independently functioning microzones, the tightly folded cerebellum contains more neurons than all the rest of the brain combined. While each region could be said to have its own form of intelligence, it's obvious none of these controls the brain as a whole.

A pinecone-shaped structure looms up ahead of us. Though only the size of a grain of rice, the pineal gland seems massive from this perspective. We slow down to regard it with quiet fascination. Not because it is the seat of consciousness, but because it was once thought to be. The great philosopher René Descartes considered this tiny endocrine gland to be the seat of the soul and all of our thoughts. That view has since been disproved many times over, despite the best efforts of modern-day would-be mystics and snake oil salesmen.

Traveling through the left lateral ventricle, we soon find ourselves traversing the corpus callosum, a thick nerve tract that connects the right and left cerebral hemispheres. This dense superhighway of axons allows for communication between the two hemispheres that would otherwise be isolated from one another.

Finally, we arrive in the prefrontal cortex. This is where so many of the functions we associate with higher intelligence and consciousness take place, including abstract thinking, learning, memory, and decision-making. We extend our search from the scale of large structures—the frontal, parietal, occipital, and temporal lobes—all the way down to that of the neurons' intracellular processes. Yet study as we might, it remains a challenge to find anything that vaguely looks or acts like it could be the brain's central command. Yes, there are clusters of neurons, glial cells, cortical columns, and so forth that perform very specific roles, each having its own function and form of limited intelligence. But none of it amounts to mission control. Each does what it has evolved to do, as it is called upon, all seemingly without conscious effort on the subject's part. At times, it feels like a vast orchestra playing an intricate symphony, without the benefit of a conductor.

We continue searching throughout the many regions of our subject's brain, from the thalamus, which helps consolidate short-term memory into long-term, to the six cortical layers of the frontopolar cortex, which is involved in holding and processing goals and multitasking. While damage or even removal of it would have some impact, it wouldn't eliminate our subject's intelligence or overall ability to function.

After millions of years of evolution, it's obvious that nearly every feature of the brain must be essential, yet no single part can answer the question: Is this the seat of intelligence or consciousness? Is this what makes humans the special beings we think we are? Which leaves us still begging the question: What makes a mind?

———

For centuries, scientists and philosophers have speculated on how the brain does what it does. For a very long time, philosophers held to a dualist belief, maintaining that the body (including the brain) and mind were distinct and separate. In the past, it was commonly believed that the intellect and consciousness (as well as the soul) were the result of some divine force and that without it, people would be nothing but empty shells. Then in the seventeenth century, René Descartes formalized this view into the mind-body problem, as it's commonly known today. For Descartes, mind and matter were two distinct aspects of nature, and one simply could not arise from the other.

A contrarian view began taking hold in the eighteenth century, with the term *monist* first being applied to the mind-body debate by the German philosopher Christian Wolff. However, it was the science of the nineteenth and twentieth centuries that really gave credence to the idea that mind arises from the body, specifically the brain.[1]

But how could something so intricate and interconnected as the brain take form in the first place? Perhaps even more perplexing, how could matter, regardless of its configuration, give rise to something as immaterial as thought, consciousness, and mind?

Beginning at its most fundamental level, the brain is a network of interconnected, intercommunicating neurons that send signals to one another via electrochemical processes that results in the transfer of electrons from one cell to another. This process originated at least once and probably many times with the evolution of voltage-gated sodium channels in early single-celled organisms. Voltage-gated ion channels are proteins that enable the passage of ions across cell membranes using either calcium, sodium, or potassium. Members of this voltage-gated superfamily are found in all domains—Archaea, Bacteria, and Eukarya—suggesting the genes for this originally appeared through lateral transmission, well before Woese's Darwinian threshold that was discussed in chapter 4.

With the appearance of voltage-gated sodium channels, rapid, relatively long-distance intercellular communication became possible. We begin to see this in the fossil record with the appearance of the Cnidarians some 580 million years ago, though some genetic analyses suggest this phylum may have originated much earlier—around 750 million years ago. Cnidarians include the hydra, whose nerve net we also discussed in chapter 4. This nerve net is distributed throughout its body, allowing a hydra to sense its environment through touch. This method of signaling in metazoans—multicellular animals—represents a major milestone in the evolutionary path to much more complex brains.

To date, the earliest evidence of a proto-brain was discovered in the fossilized brain tissue of a panarthropoda from 521 million years ago during the lower Cambrian era. This clade[2], which includes Onychophora (velvet worms), tardigrades, and arthropods, is the most numerous on Earth, with more than a million species. But because soft features such as nervous systems are challenging to preserve in fossils, we should note that they probably got their start considerably earlier than this, perhaps around 540 million years ago.[3]

The appearance of the first vertebrates 525 million years ago is further evidence of primitive brains, since one purpose of the vertebrae was to protect the spinal cord, a major pathway for communicating information from the brain to the rest of the peripheral nervous system. Again, fossils are the earliest physical record we have and very likely don't represent the first appearance of a given species. Using a method evolutionary biologists figuratively call a molecular clock, estimates of species divergence can be made through the analysis of genetic changes over time. However, this measurement is frequently disputed when looking so far back, in part because the molecular clock progresses at varying rates for different species.

Continuing our story, the vertebrates evolved into the bony fish, the amphibians, the reptiles, the mammals, and so forth. Looking at the ongoing evolution of the brain throughout this progression, we find a great many of the genes and phylogenetic traits being conserved because, as they say, success builds upon success. The medulla, for instance, probably first evolved in early fish over 500 million years ago during the Cambrian era. The basal ganglia are a group of cortical structures deep in the brain of all vertebrates, from hagfish to *Homo sapiens*. The limbic system, which is involved in emotion, motivation, memory, and learning, emerged in the first mammals. The cerebrum or neocortex begins to appear in mammals about 200 million years ago, developing in size and complexity across a range of species until we reach the primates and eventually ourselves. Evidently, when evolution finds a good thing, it likes to hang onto it.

A fascinating feature of these neural structures is that while the brain stem and limbic system are based on nuclei (dense clusters of neurons that perform a specific function), the cerebellum and cerebrum have a cortical architecture. In the latter organization, neurons are arranged into layers near the surface of the brain—the cortex—with axons (nerve fibers) traveling between those layers, but mostly they amass below them. This arrangement allows for a far greater number of connections as a function of the surface area of the cortex. Such an arrangement can benefit from maximizing surface area, the reason for the extensive convolutions that are formed throughout these two parts of later mammalian brains, particularly the cerebral cortex. This increased brain folding typically correlates to an increase in intelligence in a species.

When comparing brain size to body mass, one might expect a reasonably linear relationship just as there is for many other organs, but that isn't the case. There's a considerable discrepancy between different species, and so a different measure was conceived, an encephalization quotient, or EQ. EQ is calculated for a species by taking the ratio of its expected brain size against its actual size using regression analysis based on a few index species. This produces a plot of how much each species deviates from the expected value and is considered a much more accurate predictor of proportional intelligence than merely comparing brain-to-body mass ratios. Perhaps one major explanation for the disparities comes from the number of brain convolutions found across species. For instance, a manatee and an orangutan each have approximately the same size brains, yet the surface of the manatee's brain is nearly smooth. In contrast, the orangutan's brain is covered in convolutions, more closely resembling that of its human cousins. This pattern repeats across the animal kingdom and is a general predictor of greater and lesser intelligent species.

That's a very broad-brush view of how nature gets from cells to nerve nets to brains. But we need much greater detail about how it all works if we are to truly understand how our brains work as well as how we might use that knowledge to further AI research.

This is the goal of several large-scale projects taking place in research labs around the world today. For instance, Switzerland's Blue Brain Project (BBP)[4] was started in June 2005 by neuroscientist Henry Markram for the purpose of building "biologically detailed digital reconstructions and simulations of the rodent, and ultimately the human brain." BBP completed a simulation of a rat's neocortical column in late 2006, refining and validating it in 2007. Understanding the neocortical column, considered the smallest functional unit of the neocortex, is essential to understanding the brain. Using an IBM Blue Gene supercomputer, the project works to build a realistic model of neurons with the hope that it will eventually help us better understand the nature of consciousness.

Building on this work, the European Union's Human Brain Project (HBP)[5] is a ten-year project that began in 2013 for similar purposes using (as yet unbuilt) exascale supercomputers.[6] Also, founded by Markram, the $1 billion program's goals range from gaining knowledge about neuroinformatics (brain data that can be used for new diagnostics and medical treatments for brain diseases) to emulating brain processes and eventually entire brains on neuromorphic supercomputers. By scanning and studying brain processes on scales ranging from molecular and cellular to macro structures, researchers hope to gain a much better understanding of how the brain works.

Another subprogram of HBP is the development of neuromorphic computers that will further advance cognitive computing,[7] taking inspiration from the brain's own neural processes. As well as fully modeling different types of neurons, there is work being done to model much larger-scale structures, from cellular-level reconstructions of cortical structures (including pyramidal human neuron models) to building models of cell signaling and inhibition cascades. While HBP has a goal of fully simulating the human brain at the cellular level in real time, it is going to take many advances in supercomputing and brain scanning to make that happen. In the meantime, a number of incremental goals are being pursued, including modeling certain brain structures and processes, including the basal ganglia, hippocampus, and protein interactions. Additionally, whole-brain mouse models are being built that draw on biological data from the Allen Mouse Brain Atlas, a highly detailed genetic, neuroanatomical map created by the Allen Institute for Brain Science in 2006. Not only smaller but much simpler, a whole-brain mouse model would require much less computing power, though it would still be very substantial.

One issue with brain modeling is that the larger the brain and the more detailed you make the model, the more computer-resource-intensive it becomes. There are 86 billion neurons in the human brain,[8] perhaps ten times as many glial cells, and thousands of regions. Simulating large structures using even generalized values takes a lot of power, but that's nothing compared to refining the model to accurately represent each and every neuron, its connectivity, and spiking behavior. Only recently has it started to be possible to do this, and that is as a direct result of the exponential growth of processing power predicted by Moore's law, along with other improvements in computing architecture.

For instance, in August 2013, Japan's RIKEN research group announced they used K, then the fourth fastest supercomputer in the world, to simulate the activity of 1.73 billion neurons and more than 10 trillion synapses, about 1 percent of a human brain[9]. The supercomputer took forty minutes to simulate one second of brain activity! Five years later, Manchester University's SpiNNaker (Spiking Neural Network Architecture) was used in a similar though much smaller test for HPB's neuromorphic computing platform. The fastest supercomputer of its kind, SpiNNaker has a novel architecture optimized to perform neuromorphic computing that emulates the brain's neurons and connections. In May 2018, a team from Jülich Research Centre, University of Manchester, the RIKEN Brain Institute, and Aachen University announced "the first full-scale simulations of a cortical microcircuit with biological timescales on SpiNNaker."[10] In other words, the scaled model was reasonably accurate and operated in real time. Composed of about 80 thousand neurons and 300 million synapses, the simulation was the largest cortex simulation to be performed on the platform. It was run on less than 1 percent of the supercomputer's full capacity and the team said they would use the test data to optimize the software to make it six times more efficient. The project's top goal is to be able to simulate the behavior of aggregates of up to a billion neurons in real time. It is also working on single board implementations for robotic control that can operate with low power consumption.

Also partnered with HBP, Heidelberg University's BrainScaleS physical model machine implements analogue electronic models of four million neurons and one billion synapses on twenty silicon wafers.[11] But where SpiNNaker emulates brain models at real time speeds, BrainScaleS is an accelerated system that operates 10,000 times faster, allowing for massively sped-up simulations. This illustrates the benefits of using neuromorphic machines that are optimized for this type of work. Between SpiNNaker and BrainScaleS, HBP can begin to access the vastly different timescales involved in learning and development, ranging from milliseconds to years.

Obviously, specialized hardware like the spiking neural networks used in these computers may be one key to overcoming the challenges of whole-brain emulation. Building on this idea, IBM announced their TrueNorth chip in 2014. TrueNorth was originally developed as part of DARPA's SyNAPSE program, which ran from late 2008 to 2015 and was contracted primarily to IBM and HRL (formerly Hughes Research Labs). Focused on developing computers that could match mammalian brains in terms of scale, function, and power consumption, SyNAPSE sought to build a system that could recreate 10 billion neurons and 100 trillion synapses, consume one kilowatt (equivalent to a toaster), and occupy less than two liters of space. While these specifications are IBM's long-term goal, TrueNorth still has a way go to get there. The chip has 5.4 billion transistors and can simulate one million neurons and a quarter billion programmable synapses. Though very limited compared with the human brain, these chips can be combined on boards that allow them to scale very well. Perhaps most importantly, they solve what is known as the "von Neumann bottleneck." Most computers are designed around a von Neumann[12] architecture, which requires accessing instructions and data from memory across a limited capacity bus that connects the processor and main memory. But TrueNorth has 4,096 neurosynaptic cores, each containing all of these—computation, communication, and memory—thus eliminating the need to continuously shuttle all of those instructions and data around. The result is an extremely power-economical chip that consumes one–ten thousandth the energy a traditional CMOS chip would require.

These projects and a good many future ones will eventually allow us to emulate the way brains process information and respond to the world. But that only takes us so far. Surely, there's so much more to our brain and mind that we shouldn't simply be able to emulate them in a computer and think they will start to think like us as a result? Shouldn't there be more to it than that?

There may well be. The Santa Fe Institute's Collective Computation Group, C4, explores the role of collective behavior, interaction, and signaling in complex systems. C4's codirector, David Krakauer, makes the argument that most current computer architectures simply don't have the intelligence at their lowest levels to come close to achieving the robustness, energy efficiency, and decentralization of the brain. Looking to the organization and intelligence of individual neurons, transistors lack the qualities necessary to give rise to an emergent behavior that is the result of a collectivized bottom-up interaction. "Emergence is not a natural concept for an engineered device," says Krakauer. "It's a consequence. A computer is only trivially a collective system. It has a lot of transistors, it encodes Boolean logic, but this isn't how collective dynamics works in evolved systems." Based on this, it may be that very different approaches will be needed if computers are ever to realize emergent behaviors comparable to mind and consciousness.

Far too often in general conversation, brain and mind are used interchangeably, but as the mind-body problem suggests, they couldn't be two more different things. A brain is an array of cells that direct and process signals. It follows the rules of physics, chemistry, and biology, using matter and energy to sense its environment and respond in ways that will maximize the likelihood of its genes being passed along to the next generation of its species.

But the human mind and intelligence that emerge from those brain processes are something else entirely. They are neither matter nor energy, yet the product of both. The mind writes sonnets. It ponders the universe. It falls in love. Our minds get nostalgic for the past, excited about the present, and dreamy about the future. At first glance, none of these activities seem like they could possibly originate from three pounds of living tissue, and yet they most certainly do. And if they can result from billions of years of evolutionary processes, then perhaps they can also be realized through directed technological means as well, given the right tools, insights, and enough time.

In recent decades, psychologists and neuroscientists such as Joseph LeDoux and Michael Gazzaniga[13] have talked about the idea of the mind's processes being relatively modular. As Gazzaniga has observed:

> The most compelling evidence for a modular brain architecture arises from the study of patients who have suffered a brain lesion. When damage occurs to localized areas of the brain, some cognitive abilities will be impaired because the network of neurons responsible for that ability no longer functions, while others remain intact, tooling along, performing flawlessly. What is so intriguing about the brain-altered patients is that no matter what their abnormality, they all seem perfectly conscious. If conscious experience depended on the smooth operations of the entire brain that shouldn't be what happens.

Similarly, Marvin Minsky began exploring something along this line in the 1970s and later in his books, *The Society of Mind*[14] and *The Emotion Machine*.[15] Minsky proposed that the mind is made up of hundreds, perhaps thousands of component processes and properties he calls "agents" (Gazzaniga refers to these as "modules"). Many agents and processes perform roles such as grasping, balancing, moving, and so forth and would have evolved as separate functions and processes over many millions of years so that we could call upon them when needed, combine them, and perform highly specific tasks. Mostly semiautonomous, they operate almost unconsciously for us. Many of the agents and processes may function as they are needed: some will be active, others less so, and some quiescent. As this interplay occurs, differing degrees of communication take place between them, and many agents may never communicate with each other at all.

Taken individually, the intelligence of these agents would be considered very limited, relatively speaking. They perform their task or subfunction, potentially with almost no conscious oversight, then hopefully get out of the way so other agents can do their jobs. This premise allows animals with far smaller and less complex brains to behave and survive without the "benefit" of consciousness. Despite our tendency to believe we control most of what we do as we go about our lives, cognitive psychologists and neurologists have determined that we are only aware of a bare fraction of what takes place in our brains.

Don't believe it? First off, consider all of the processes you acknowledge you are unconscious of nearly all of the time, unless something goes wrong with them. For example, the autonomic nervous system, which is made up of the sympathetic, parasympathetic, and enteric processes, handles everything from our heart rate, respiration, pupil response, and digestion to coughing, sneezing, swallowing, and vomiting to sexual arousal and orgasm. Then there's our endocrine system, its glands secreting hormones—messenger chemicals—that regulate our body's organs and help us maintain homeostasis (physiological equilibrium). These often influence our emotional state, which in turn is also capable of further influencing our endocrine system. While we can take actions that affect many of these, for the most part they respond entirely without conscious effort on our part.

Now consider that you're sitting at your computer drinking a cup of coffee. As you read the morning news, you pick up your cup, perhaps without even looking at it. Are you aware of the muscles you're activating to grasp the cup, lift it, and guide it to your lips? Perhaps you taste the coffee and think for an instant that it's a different blend than you usually drink. Perhaps not. But unless it's scalding hot or disgustingly cold, you probably don't give its temperature much thought. Do you deliberately activate your taste buds or heat receptors to communicate any of this to your brain? What about all of the muscles needed to direct the warm liquid to the back of your throat? Do you will your nasopharynx closed, equalize the pressure between it and your middle ear by closing your auditory tube, use your palatoglossus to close your oropharynx, close your larynx when your uvula and other receptor nerves tell you it's time, sequentially contract your superior, middle, and inferior pharyngeal constrictor muscles, and propel the liquid into your esophagus? (I'm really glad we don't need to be aware of this process. I'm exhausted just thinking about it!)

Or you go into the office, and the first thing you have to do is attend a department meeting. You grab your agenda and papers, head to the meeting room, and take your usual seat.

How many steps did you take between your office and the meeting room? It doesn't matter, since you very likely don't remember walking there at all.

How much of our daily activity are we unaware of? Many cognitive psychologists put our unconscious activity at 95 percent, while others suggest it may be as much as 99 percent. The actions, behaviors, and decisions we are truly aware of make up only a tiny amount of our brain's actual output.

This is where things get interesting. Take a moment to stop reading and just listen to your environment. What do you hear? An appliance running? A dog barking relentlessly outside? A background conversation you're only now becoming aware of? Whatever it is, why have you only become aware of it just now? The sound was present, it was activating the cilia and nerve cells in your inner ear, and it obviously wasn't below your threshold of hearing because you can hear it now.

So why wasn't it part of your consciousness immediately before this? Why aren't we conscious of everything around us all of the time?

If we consider our brains as a collection of agents or modules that become active and inactive according to our needs of the moment, these behaviors begin to make a lot more sense. Of course, much of that was about controlling our bodies, but the same concepts could give rise to various aspects of consciousness as well. Perhaps at some point in our history, our ancestors' emotional and social agents started activating in such a way as to result in cognitive empathy, putting us in the shoes of others. In turn, this caused us to think differently about ourselves and other people, leading eventually to theory of mind, our ability to model what other people are thinking and feeling.

Other processes that were used to observe our environment, and especially our observations about others (skills key to our success as an increasingly social species), would be turned inward over time, observing our own emotions, thoughts, and other mental states. As language agents developed, they facilitated the acquisition and use of concepts that in turn furthered self-talk and internal dialogues, the stories we tell ourselves.

Along the way, perhaps the many agents integrated so seamlessly and responded to each other so instantaneously that our sensory agents blended with our cognitive and emotional ones, allowing us to perceive the world, allowing us to be overcome by the beauty of a sunset. To thrill at the trill of a finch. To be transported by the scent of a pine tree. Experiencing all of the world as we never had before. Theory of mind, self-awareness, metacognition, subjective experience, and even phenomenal consciousness may all be the result of this intricate fusion.

Will artificial intelligence realize a similar aggregation of functions and modules? Or is there something so fundamentally different between biological and electronic processes that this could never be possible? On the other hand, is the ongoing disagreement over AI's connectionist and symbolic approaches a false dichotomy? Certainly, we human beings rely on symbolic systems, especially since our ascendance from the early hominids. There's no denying that. But if this ability is the result of particular modules that evolved to better allow the manipulation of symbols and logic, then they are just one more instrument in our cognitive tool kit. The underlying mechanisms implementing this symbolic cognition are inarguably still connectionist. Every part of the human brain, from our visual system to memory formation, storage, and retrieval to abstract thinking begins from connectionist foundations. Increasingly advanced AIs, including artificial general intelligence, should be no different once the necessary fundamentals are in place. While the early symbolic focus of Good Old-Fashioned AI was an essential step in uncovering the path AI is currently on, it feels as though we were over-abstracting and beginning at too high a level. Just as with biology, a low-level connectionist underpinning may be required before the higher-level abstractions can properly develop.

The connectionist interpretation appears to be supported by *network neuroscience*. This relatively new field takes a perspective informed by complexity science and combines it with mathematical tools, such as graph theory, in which the complex connections of the brain are represented as graphs made up of nodes and links. From doing this, all sorts of new insights are being gained into the many levels of networks that make up the brain. As Provost Professor in Psychological and Brain Sciences at Indiana University Olaf Sporns has noted: "The topic of networks in the brain covers very different levels of scale—all the way from social networks and networks of individuals as they engage in collective behavior, all the way down to the kinds of molecular, genetic, regulatory networks that we see engaged in development and other processes in cells."[16]

Whether talking about billions of neurons acting in concert or the interaction of hundreds of cognitive modules, what we have here are many highly similar individual units operating together in direct, near-instantaneous communication with several of their immediate and sometimes not so immediate neighbors. These units typically follow a set of relatively simple rules defining how they will interact relative to one another. In the case of the human brain, this apparently resulted in generating the novel emergent properties we call cognition and consciousness, phenomena that are most certainly greater than the sum of their parts.

Sound familiar? Like the termites building towering cathedral-like mounds and the murmurations of starlings moving as a single entity, a multilevel neural complex following a relatively simple set of rules yields emergences that increase its future freedom of action, in a way that cannot be predicted from its constituent parts. Once again, the vortexes and eddies of the universe result in little islands of complexity and emergent behavior, temporarily sequestered from the vast entropic force of the cosmos.

What does this mean for the future of technological intelligence? Unfortunately, the very nature of complexity and emergence precludes our anticipating the exact nature of a highly complex emergent process. After all, as we saw with Reynold's boids simulation and Wolfram's automata, even the application of a few simple rules can lead to behaviors we cannot accurately predict. How much more challenging then to anticipate the nature of a mind that originates from countless neuromorphic circuits, not to mention a mass of billions of interconnected brain cells?

Yet if we are simulating those same brain cells and structures *in silico*, shouldn't we be able to expect a similar emergence? Shouldn't that be all it takes to make a mind?

Perhaps. But ask yourself: how accurate is the simulation? Even if we are able to reproduce a brain perfectly based on our current knowledge and understanding, that doesn't mean the reproduction would actually be perfect. At best, it will only be as good as the state of knowledge at that moment, which in large part is only as good as the current tools we bring to bear on the problem. If the spatial and temporal resolutions of the equipment we use to scan the brain fall short of some as yet unknown threshold, then such a hoped-for emergence may never even happen.

Of course, it is always possible that this low-fidelity configuration will still yield some form of emergence, though it would probably be very different from the human mind we're trying to model. It's a long shot, but it can't be ruled out.

So here we are, faced with yet another limitation of the emulation approach, which suggests that perhaps the best strategy might be to work with an AI's default native state. We already have neural networks of every flavor and configuration that can do all kinds of unexpected and amazing things. In many respects, the functions of artificial neural networks are themselves emergent behavior. Yes, it's trainable emergent behavior, but that's essentially what our brains—our own personal neural networks—do.

If we were to demand that our own neurons combine and interact in ways other than what they are optimized for, would we still experience the emergent behaviors we know and love? I'm guessing not. So why anticipate or even expect this of systems that attempt to replicate our biological processes suboptimally? Better to design them to operate from their strengths rather than hobble them trying to emulate ours.

Our own brains are made up of an amazing array of repeating, interlinked structures, from nerve cell clusters to cortical columns and pyramidal neuron architectures. Our genes do not explicitly tell our brains to grow and locate these individually, but rather to iteratively replicate them into structures within certain boundary conditions. From there, the still-being-discovered rules of neurogenesis propagate and prune the connections between the neurons as we go about our business of learning about the world. Or, as neuroscientist Michael Gazzaniga has put it: "The brain's large-scale plan is genetic, but connections at the local level are activity dependent and a function of epigenetic factors and experience: Both nature and nurture are important."[17]

Approached from this extensive and complex perspective, we cannot expect to directly implement all of the agents an AI will need in order for it to become intelligent, self-aware, or conscious. There are simply too many specifics to design, and the process would be too complex. But fortunately, it's very likely we won't need to. Work is being done on automated machine learning, or AutoML, that can select and assemble the best algorithm or neural network for the task at hand. For instance, at Google, researchers have used what's known as "reinforcement learning" to automate the machine learning development process. In one test, they directed their AutoML program to build a "child AI" that could recognize objects in a real-time video. The AI-bred AI beat out all of the AIs designed by human developers. The system was 82.7 percent accurate on the ImageNet validation set, more than 1 percent better than any previous results achieved by human developers.

The ability of an AI to select the best algorithm or neural network on the fly and then to optimally tune itself has enormous potential. Imagine being able to "grow" different AIs based on a range of agents and functions that are not unlike the modularization of our own biological brains. Relying on machine learning optimization, such an approach could lead to all manner of insights and breakthroughs. Techniques like AutoML could select the best method for implementing any given module. Alternately, or more probably in conjunction with this scheme, AI "brains" could implement a new module or agent on the fly, growing additional resources on an as-needed basis.

Of course, we can't ignore that we will soon be entering very new ethical territory. If these AIs become increasingly self-aware or otherwise conscious, we may reach the point that we could be subjecting them to psychological stress or even pain. As ethical beings, we cannot allow this to happen and will have to find ways to prevent it. In and of itself, this problem will be sufficient to fuel an entire field of study.

Finally, the capacity for self-design and recursive self-improvement would bring us dangerously close to the conditions many people are concerned will lead to an intelligence explosion, sometimes referred to as the *technological singularity*. There are a number of reasons the technological singularity may and may not be possible, and we'll investigate both issues in chapters 13 and 14.

In many ways and to the best of our knowledge, our brains and minds are the apex of emergent information processing, at least in this tiny corner of the universe. However, we now find ourselves on the cusp of new emergences that may soon rival our own. These emergences may go on to develop and evolve along their own unique course, or in time they may become integrated with our own intelligence, resulting in something even more unexpected. Then again, perhaps both paths will be taken, as we will continue to explore throughout the rest of this book.

CHAPTER 11

THE PATH TO INTELLIGENCE AUGMENTATION

"There are no hard problems, only problems that are hard to a certain level of intelligence. Move the smallest bit upwards, and some problems will suddenly move from 'impossible' to 'obvious.' Move a substantial degree upwards, and all of them will become obvious."

—Eliezer Yudkowsky, AI theorist and researcher

Isaac entered the tiny consumer electronics store a little too quickly, nearly running into the back wall as the door closed behind him. He took a quick step back. Really, to say this store was tiny was a major understatement. To be accurate, it was more of a booth or a closet, and a spartan one at that. The smooth ice-white walls were completely devoid of shelving or signage, save for the familiar cool blue emBrain logo centered on the wall in front of him. For an instant, he thought the logo was shrinking, until he realized the wall was steadily receding from him, nearly doubling the size of the booth-closet-room.

"How may I help you?" asked a disembodied voice that subsequently became embodied. Before him stood an attractive young woman he knew wasn't a woman. This was the store's avatar, a virtual incarnation of its AI.

"Hi," Isaac responded, trying not to look too startled. "I wanted to find out about an upgrade."

"Do tell," replied the avatar sarcastically, cocking her head to one side. It was a response many customers would've thought annoying, but Isaac found it oddly appealing. He had to remind himself this wasn't an actual person he was dealing with, but a highly intuitive sales program.

"Well, welcome to emBrain, the only other brain you'll ever need," the avatar said, quoting the company's trademark tagline. "You can call me Trish."

"And you know I'm here for an upgrade because it's the only thing you sell."

"Exactly," Trish the avatar replied. "I see it's been a while." She peered at Isaac as though she was looking directly into his skull. "Wow! You're still on BCS2057? You like to live dangerously."

Isaac shrugged, a little embarrassed. The year was 2062, and he knew he'd held off upgrading his BCI—brain computer interface—for way too long. "I've been saving up," he said, rationalizing the delay. "Anyway, I wanted to give them a chance to get the bugs out of the latest version."

"Seriously?" Trish said incredulously, holding her thumb and forefinger millimeters apart. "You're this far from a head full of serious data corruption."

"How do you know that?" he asked suspiciously. He knew full well his headgear had been acting up, getting increasingly quirky.

"The same way I know you didn't see the terms and conditions when you walked through the door. The one that authorized my running a full diagnostic on your headgear, Isaac Mendez of Salem, Oregon." Pausing dramatically, the avatar added, "Let me show you what you're missing."

The empty ice-white room suddenly expanded to the horizon, transforming into a lush tropical jungle filled with a menagerie of exotic creatures. A scarlet kingsnake with the face of a famous pop star slithered across the top of Isaac's left foot. Without any effort on his part, Isaac immediately identified the snake as *Lampropeltis elapsoides*, a nonvenomous species. The face of the popstar belonged to Flavio, a neurobeat sensation from two years earlier. His entire discography instantly appeared in Isaac's head. With a thought, Isaac sent one of the tunes emanating throughout the jungle. He turned full around, his mouth gaping at how much he suddenly knew about anything, everything. He turned back to Trish just as the sky above them abruptly filled with flocks of flying monkeys and iridescent dragonflies the size of helicopters.

"Wha—. . . wow . . . That's crazy real!" Isaac exclaimed.

Trish the avatar smiled. "BCIs have gotten a lot better since that junk you're using came out. Your processor can't do a fraction of what today's headgear can do. But most important are the connections. Our new emBrain CogLink has full neuron-level real-time resolution with direct and augmented feedback enhancement. As far as your brain is concerned, any input from it is indistinguishable from reality."

"Whoa! I can even smell the jungle!"

"Sight. Sound. Taste. Touch. Smell. Our latest version interfaces directly with your visual, auditory, gustatory, somatosensory, and olfactory cortexes. It's all piped directly to your brain."

"But I haven't even upgraded yet. How am I experiencing any of this?"

"We 'jacked your headgear when you walked in, and now we're feeding interpolated signals to your existing equipment. Again, you agreed when you entered the store. You didn't see our T&C because your gear is so antiquated. No, four years old is antiquated. Your five-year-old junk? That's ancient."

Isaac barely heard the avatar, caught up as he was in the fully immersive illusion of *neureality*.

"Because of that," Trish continued, "this isn't nearly the full experience you get with BCS2062, but obviously it's way better than what you've been using." A miniature amber-colored flying dragon

swooped down and landed, perching on Trish's shoulder as she continued her patter. "This hi-rez emBrain experience can be yours for
twenty-four easy semimonthly payments. Or you can go back to the
way things were before you walked through that door."

Isaac didn't need to think twice to know his answer.

"Where do I sign?"

While brain computer interfaces have been a standard trope of science
fiction novels and movies for generations, today they are beginning to
feel more like an inevitability. The power to read thoughts, to communicate telepathically, and to control matter with nothing but our minds
seems like comic book fantasy, yet these abilities are already here.

Of course, a phrase such as "with nothing but our minds" can be
very misleading. A considerable amount of high technology is needed
as an intermediary between the user and the task at hand. Nevertheless,
the day is fast approaching when these BCIs will be the preferred and
dominant method for controlling and directing our technologies.

These devices are a direct continuation of the ongoing evolution
of user interfaces we discussed earlier. This has been the progression as
people have sought to make technology, and computers in particular,
more accessible and more capable of working on our terms. From
hardwired instructions to punch cards and punch tape, to command
line interfaces, to graphic user interfaces combining a monitor, keyboard, and mouse, to natural user interfaces (NUIs), we have made
the controls more and more human-friendly. With NUIs, we have
reached the stage where we are commanding and communicating
with computers and other devices much as if they were people, using
speech, gesture, touch, and even emotions. The trend has resulted in
the democratization of such technologies, making them accessible to
users with little training, and also opening enormous markets to be
capitalized on. Without touchscreens, for instance, smartphones would
be little more than clunky toys. Without speech recognition, a rapidly
growing number of devices could not be hands-free.

Now we find ourselves on the cusp of an interface that is like nothing we have ever seen before. At the same time, it will eventually become the most natural means we have for communicating, interacting, and controlling our world with our minds. Our mind has always been what initiates all of our actions, our direct connection to the world, while our bodies have been the intermediary, the means by which we enact those thoughts through language and other tools. With BCIs, we now have the first means for engaging our tools and thus our will without any intermediary actions on the part of our bodies.

This technology marks the beginning of a new era for human intelligence, offering a long-term potential for rapidly augmenting, improving, and accelerating our intellects. Like the inventions of language, the printing press, libraries, and the internet, BCIs will enhance what we can know and how we will know it. They will make it possible for us to be vastly smarter, collaborate in new ways, and access resources previously unheard of. (This is not to suggest that they will necessarily make us any wiser, though they may yet surprise us and do that too.)

What are these devices that are set to make gods of humanity? How far have we come to date, and how far do we have yet to go? More importantly, what are the benefits and opportunities, the perils and pitfalls of this transformative technology?

Brain computer interfaces, sometimes called brain machine interfaces or neural control interfaces, consist of a number of different technologies that detect neural activity in the brain. These range from noninvasive devices that generally are placed in contact with the scalp, to partially invasive BCIs that are placed inside the skull, directly on the surface of the brain without penetrating it, to invasive BCIs in which electrodes are actually implanted in the brain. Each has its strengths and weaknesses. Generally speaking, invasive BCIs result in the best, most accurate signals due to their being in direct contact with the neurons, but carry the higher risks involved with any surgical procedure. Additionally, the body's immune system attacks the electrodes and points of contact, leading to corroded electrodes and/ or scar tissue, which greatly reduces signal quality over time.

One of the earliest invasive BCIs was Cyberkinetics' BrainGate, a technology that uses an array of thin electrodes to penetrate the brain's cortex. BrainGate's first major success was achieved in 2004 when tetraplegic Matt Nagle was fitted with it, allowing him to operate a cursor on a screen, as well as the first mentally controlled robotic arm.[1] The ninety-six-electrode array inserted into his brain detected neural activity and converted it into the control signals required to operate the robotic arm.[2]

Years earlier, the first efforts to explore BCIs had been funded by DARPA and involved finding usable controllable signals in EEG (electroencephalography) data, a technology that had been used for decades in the medical profession. While EEG typically generates lower quality, lower resolution signals than implants, its noninvasive nature makes it far more suitable for general and especially amateur use. As EEG technology and computing have both progressed, what can be achieved with it has also advanced, so that today EEG BCIs have been used for gaming, for robotic wheelchair control, to operate prosthetics, for one- and two-way communication, and by amateur developers.

In addition to EEG, a number of other neuroimaging techniques have been used as noninvasive BCIs, including functional magnetic resonance imaging (fMRI), positron emission tomography (PET), and magnetoencephalography (MEG). Each has its pluses and minuses in spatial and temporal resolution, portability, and cost, and therefore some are better suited to one type of application or research use than another.

Somewhere between invasive and noninvasive are the partially invasive methods. Perhaps the best known of these uses electrocorticography, or ECoG, in which sensors are placed in direct contact with the brain without penetrating it. This method can provide higher spatial resolution, better signal-to-noise ratio, and wider frequency range than EEG, while having lower risks of scar tissue and other health issues caused by implanted electrodes. The potential for partially invasive BCI is likely to grow as advances are made in a number of other fields, including materials science, biotechnology, nanotechnology, and power harvesting.

Of course, there are many ways to increase and improve our brain's ability to learn and focus. Nootropic drugs such as Adderall, Ritalin, and Noopept have been shown to provide a temporary boost in attention, information processing speed, and memory formation and retrieval. Neurostimulation techniques such as transcranial magnetic stimulation (TMS) and transcranial direct current stimulation (tDCS) can also improve attention and learning ability according to some research. TMS uses magnetism to alter the activity of neurons, and tDCS uses electricity. However, the exact mechanism by which these methods appear to strengthen or weaken synaptic transmission between neurons remains uncertain. While all of these methods comprise important areas of focus and research, they appear to be altering neurotransmitter behavior, "pushing" otherwise unmodified neurons beyond their natural operating conditions. Just as overclocking a CPU past its standard specifications can lead to problems, overdriving our brain cells could potentially cause temporary or permanent damage. For these reasons, I would suggest that BCIs will eventually offer a longer-term means of not only boosting but of monitoring and otherwise augmenting our thought processes by off-loading much of the additional workload to other nonbiological resources.

It is safe to say that we are still only at the beginning. The nearer-term possibilities for BCIs are enormous, with the potential to be used for therapeutic purposes, including the recovery of mobility through the control of robotic prosthetics and wheelchairs as well as communication methods for those with locked-in syndrome. While technologies such as artificial retinas for restoring vision and cochlear implants that restore hearing are less formally tied to these interfaces, what is learned from work with them will be important to our understanding of integrating our senses with these nonbiological systems.

Beyond this, in time we will see advances in our abilities to not only interact with our computers and the internet but address and work with our increasingly smart environments. The Internet of Things, which is vastly increasing connectivity throughout our world, is seen as providing tremendous opportunities for BCIs because of the many possible ways it could be used to empower users. Last, but certainly not least, the longer-term ability to rapidly access and acquire knowledge as well as integrate on-the-fly computer processing power could see humanity leapfrog what would otherwise take natural evolution millions of years to accomplish. But to get to that point, many advances will still need to be made. Perhaps most significant is two-way neural communication.

Just consider how far this technology has come in only a few short decades. From the early investigations in the 1970s into the viability of using EEG, we began seeing useful advances and applications less than a quarter century later. Within only a few decades, BCIs would be used to write and send tweets on Twitter, this being achieved at the University of Wisconsin–Madison in 2009.[3] In 2014, a team of researchers transmitted the words "ciao" and "hola" from the brain of one person in India into the brain of another person in France, five thousand miles away.[4] At the University of Washington in 2013, two players engaged in a game in which one person mentally issued a command to press a button that activated the hand motion of a second person in another building—without that person consciously willing their own hand to move.[5] And by the middle of that decade, a growing number of researchers were finding ways to read and identify specific words as their test subjects thought about them.

As a serious control interface, being able to use words is critical. Language is the means by which we manipulate concepts and communicate with each other. Without it, nuance, interaction, and especially interpersonal communication will be difficult at best. That's why MIT Media Lab's AlterEgo, demonstrated by its inventor, Arnav Kapur, in 2018, is such an important development.[6] Using an adhesive strip with embedded electrodes, it picks up neuromuscular signals along the throat and jaw that are activated when the user thinks about a particular word. The user can then hear responses from the device through bone conduction. Similar to earlier subvocalization techniques[7] (subvocalization is thinking about speaking), AlterEgo claims to identify words correctly 92 percent of the time. While this early prototype uses external electrodes to detect the nerve signals, a later version could presumably do the same using a subcutaneous detector. Additionally, while this is not a true BCI, the integration and use of the peripheral nervous system provides a direct, low-risk conduit to and from brain. Whether detected directly in the brain or via a nerve nine inches away, the result is much the same.

A series of 2018 papers from three different teams[8] suggest that such direct language reading using EEG BCIs may not be that far off either. The goal of all of these projects was to develop a way for people who have lost the power to produce speech (such as through ALS, amyotrophic lateral sclerosis) to regain that ability. Some of the projects used deep learning that was trained on test subjects as they read words that activated the firing of neurons specific to that speech.

Then in 2019, researchers at Columbia University's Zuckerman Institute succeeded in translating brain signals directly into speech.[9] Using ECoG (electrocorticography), they detected the brain responses as the subject thought specific words.[10] These signals were then processed through a deep neural network, which was subsequently fed into a speech vocoder (voice encoder/decoder). Though distinctly synthesized, the results were intelligible. According to the study's lead scientist, Nima Mesgarani: "In this scenario, if the wearer thinks, 'I need a glass of water,' our system could take the brain signals generated by that thought, and turn them into synthesized, verbal speech." Obviously, the potential of this technology to transform the lives of those who have lost the power of speech is tremendous. But it also marks another milestone on the path to translating thoughts directly into words.

It's still very early days, but these achievements point to a future— probably a relatively near future—in which we will increasingly integrate BCI technologies with our brains. If we look at the course of history and the technologies that already augment our intelligence and access to knowledge—books, telecommunications, the inter-net—they have trended toward being more immediate and accessible over time. Today, our ever-present smartphones are our supplemental brains in everything but name, giving us virtually instant access to nearly all of the world's knowledge. We may be little more than a generation away from having the capability to tie these directly into our nervous systems on a daily basis. An artificial external brain, an *exocortex*, could augment our own natural intelligence, doubling it, then doubling it again in a virtuous cycle that gives rise to ever greater intelligence.

What will it take to reach this level of integration? A good many researchers and entrepreneurs have asked themselves the same question, as have a number of government agencies. Inspired by the Human Genome Project, in April 2013 the Obama administration announced the Brain Research through Advancing Innovative Neurotechnologies (BRAIN) Initiative.[11] Its goal is to revolutionize our understanding of the human brain through the development and application of innovative technologies in order to better treat neurological and psychiatric disorders. The initiative, which funds a wide range of research across the sciences, seeks better methods of scanning and recording the brain's activities as well as improving our understanding of the underlying processes taking place within it.

Among those working with the initiative, DARPA and IARPA (Intelligence Advanced Research and Projects Agency) have multiple projects focused on several different concerns. For instance, the military must deal with millions of soldiers suffering from post-traumatic stress disorder (PTSD) and other neuropsychiatric illnesses. It is estimated that 20 to 30 percent of soldiers have some degree of PTSD, which correlates strongly with the number of deployments they have served. Because of this, DARPA's SUBNETS[12] program (Systems-Based Neurotechnology for Emerging Therapies) was launched to develop a brain chip that not only can detect and record brain signals associated with PTSD, but will also use that information to send appropriate stimulation to the affected neurons in order to break the neurological cycle that contributes to the disorder.[13] Such a "closed-loop system" holds enormous promise not only for therapeutic purposes but also for learning more about how and why the brain works the way it does.

Another DARPA program designed to help returning soldiers is RAM (Restoring Active Memory),[14] a fully implantable neural device designed to restore episodic memory function in those suffering from TBI, or traumatic brain injury. Guided by studies of the hippocampus, a region of the brain critical to memory function, this work also uses a closed-loop system, monitoring neuron states and issuing appropriate stimulation in order to activate and enhance memory formation and retention. The work could eventually lead to an implantable, artificial hippocampus, a neural prosthetic designed to improve memory function in those with TBI or Alzheimer's. Beyond this, it has been speculated that this research could also one day lead to a method for enhancing existing healthy brains as a method of supercharging memory.

IARPA has several programs designed with the purpose of keeping the US at the forefront of intelligence work. Their SHARP program (Strengthening Human Adaptive Reasoning and Problem Solving) seeks to improve human reasoning and problem-solving skills, while MICrONS (Machine Intelligence from Cortical Networks) is working to reverse engineer the human brain in order to improve machine learning. KRNS (Knowledge Representation in Neural Systems) is developing and evaluating theories of how humans represent conceptual knowledge. ICArUS (Integrated Cognitive Neuroscience Architectures for Understanding Sensemaking) wants to understand our ability to detect patterns in data and then infer underlying causes in those patterns. Developing cognitive models, the project will then implement them in software for the purpose of bringing the ability to artificial intelligence.

From the foregoing, it should be evident that these agencies are striving to advance and stay at the forefront of AI and IA (intelligence augmentation)—as well as having a love for tortured acronyms.

In all of this research, we can see that a large part of this work involves decoding the language of the brain. How the firing of specific neurons translates thoughts into spoken words or results in retrieving a specific memory remains a mystery. Having an improved ability to read and interpret the interplay between neurons would advance our understanding of the brain significantly. Fortunately, such a tool and ability does exist, and it was developed relatively recently.

Optogenetics is a biological technique that makes it possible to monitor and control neuron states in near real time. Combining optics and genetics, genetically modified neurons become capable of emitting light when they fire, using genetically encoded calcium indicators (GECIs). These fluorescent molecules emit a tiny flash of light from a voltage-sensitive fluorescent protein when they fire, which allows the monitoring of the calcium channel status—part of the neuron's activation sequence.

Using another set of genetically encoded light-sensitive compounds called "opsins," originally derived from microbial organisms such as green algae, neurons are triggered or inhibited in a highly controllable manner. When flashed with laser light guided by fiber optics, the light-gated ion channels in the modified neurons are activated. This makes it possible to alter the neuron's spiking behavior in real time, far more rapidly than using other techniques. Lab tests on animals have demonstrated the ability to alter locomotion, feeding, stress resilience, and sexual behavior in subjects with this technique.

The potential of optogenetics is tremendous, initially for the study of the brains of animals and later in humans. Already numerous studies exist focused on using this technology for therapeutic purposes. In time, it may even play a huge role in developing its own form of BCI.

This tool could be extremely important, because as neuroscientists such as David Eagleman of Stanford University and Blake Richards of University of Toronto have pointed out, electively implanting chips and neural prosthetics like BCIs in healthy brains is highly unlikely not only in the near term but for the foreseeable future as well. The risks of infection, brain damage, and other considerations will make it an ethical nightmare for many years to come.

Nevertheless, research that eventually results in new therapeutic approaches will certainly continue, propelling our knowledge of innovative techniques until one day they will, in all likelihood, be applied in healthy individuals too. As Eagleman observes, "I do think that related approaches are more feasible: from internal solutions such as neural dust, genetic approaches, nanorobotics to next-generation technologies of high-resolution measurement and stimulation."[15]

This doesn't mean that work on implantable electrode BCIs for intelligence augmentation will stop, however. As mentioned, DARPA and IARPA are hard at work advancing the field in search of new therapies, but several of the programs also indicate designs to enhance the existing abilities of their soldiers and agents.

Then there are high-tech startups like Kernel that are on similar paths. Founded in 2016 and funded by Bryan Johnson, formerly of Braintree (which he sold to PayPal for $800 million), Kernel is focused on developing BCIs and other techniques by applying Silicon Valley–style approaches to the problem. As Johnson states, "I think that unlocking the brain and learning how to read and write our neural code is the single most consequential and exciting adventure in the history of the human race."[16]

The company is guided by a team of advisers that include Ed Boyden, one of the inventors of optogenetics, and the aforementioned David Eagleman. Initially, the company stated they weren't interested in focusing on one specific disease or device so much as building a broad technology platform. According to Johnson, "A multi-product approach, staged over multiple years, optimizes for expertise accumulation and technological and scientific breakthroughs."[17] However, more recently the company has turned its attention to what will likely be more immediately attainable goals, such as developing an artificial hippocampus to treat memory disorders, including those caused by TBI and stroke.

Another company that has set its sights on the future of BCIs is Neuralink, founded in 2016 by entrepreneur and engineer Elon

Musk. Like Kernel, Neuralink has an initial goal of being able to treat brain impairments, such as those caused by stroke or external trauma. However, the secretive company has acknowledged that its eventual goals are much larger. Ultimately, it plans to develop a form of implantable device that will allow a healthy brain to access external knowledge, whether from a hard drive or the internet, at broadband speeds.

Musk, like Johnson, maintains that we are on a very dangerous course with artificial intelligence and will see humanity rapidly surpassed unless we take steps to prevent it. The approach these two speculative pioneers envisage is to augment the intelligence of human beings so that we might better keep pace with the silicon superintelligences that await us in the future. Then eventually, either as an ongoing progression or perhaps a strategy of last resort, we might use these interfaces to merge with the technological superintelligence or superintelligences, as the case may be. This fusing of human and artificial intelligence would theoretically raise humanity up, taking it to what would be its next stage of evolution.

Musk referenced NeuraLace repeatedly in the early days of the company. A concept from the science fiction writings of Iain M. Banks, NeuraLace is envisaged as an ultrathin mesh implanted around and through the brain. Its presumably vast array of electrodes would allow monitoring of the brain at the neuron level as well as writing to it using appropriate levels and configurations of electrical stimulation. Such an ability to read and write information to and from the brain could completely transform the human species. While criticisms have been raised about losing our humanity and becoming cyborgs, Musk points out we are already well on our way to transcending our biological origins. Our reliance on our personal supercomputers—the smartphones in our purses and pockets—already makes us many times smarter than we would be without them. Though it is questionable if this instant access to knowledge is the same as making us smarter, it is difficult to argue with the notion that we are becoming more immediately interdependent on our technology.

Numerous other companies are using different approaches that are at varying stages of development. Social media giant Facebook has been at work on their own BCI for several years, developed by their secretive Building 8 division as part of their Silent Voice First project. The goal for the BCI is to "read" the words a user silently speaks to themselves, allowing them to type up to a hundred words per minute using nothing but their thoughts.

The startup CTRL-Labs uses two different methods, which they call myo-control and neuro-control, in their efforts to build a BCI. One method takes its cue from the nerves involved in muscle activation, while the other listens to specific neurons in the spinal cord. Based on their research, the company has developed a wearable wristband that allows users to control devices by simply thinking about moving their fingers and hands. In September 2019, CTRL-Labs was acquired by Facebook for an amount thought to be as much as one billion dollars. The startup will join Facebook's Reality Labs group, which works on virtual and augmented reality products.

There are probably hundreds of ventures working on these problems around the world. One open source community-based approach is OpenBCI. Originating from a 2013 Kickstarter campaign, the organization produces circuit boards that can measure and record brain activity either via EEG for the brain, EMG for the muscles, or EKG for the heart. OpenBCI also makes design files available for a 3-D printed headset called Ultracortex.

Beyond this, a number of futuristic approaches for developing BCIs have been proposed. *Neural dust* is a term that has been applied to the ever-shrinking sensors and devices that could be used one day in lieu of electrodes. In 2016, a team of engineers from UC Berkeley built the first dust-sized, wireless sensors that can be implanted in the body. These batteryless microsensors could one day be miniaturized far below the millimeter scale, transmitting electrophysiological states from inside the brain. On the heels of this, in 2018, the same team of Berkeley researchers announced StimDust, a micro-nerve stimulator several times smaller than a grain of rice. As materials science advances and such devices continue to shrink, the potential for this partially invasive method will grow.

Continuing on the course of ever shrinking technology, various ideas about nanorobots, otherwise known as nanites, have been speculated on as well. Small enough to enter a living cell, millions if not billions of theoretically sub-neuron-sized nanodevices could read, write, and collectively interact from anywhere within the brain. Though still many decades away, our advancing ability to control matter at the molecular level suggests that one day nanites will be within our ability to achieve as well.

All of this effort indicates a major awareness and interest in this nascent but rapidly accelerating field. Who will be the winners when the neural dust finally settles, it's impossible to say, but one thing does seem certain: we will be using our brains to directly communicate with our technology in the future.

But what if that's not the future some of us want? What options will we have if we choose to opt out of the great BCI race? Unfortunately, there may be very few. A future filled with seriously intelligence-enhanced coworkers and rivals is not going to be particularly sympathetic or benevolent to any unimproved predecessors. The competitive nature of our world may ensure that those who opt out, whether from principle or due to economic inability, will be quickly cast aside. The detritus of the brave neural world, they and their progeny could be forever lost, dead-end limbs on evolution's tree of life. Consequently, we might expect that this transition will result in considerable resistance and conflict along the way, a situation we would be wise to try to avoid.

The story of technology has been one of continual change, in which those on the wrong side of history have routinely been left in the dust. The primary lesson of history has been that while it is possible to influence the direction of progress, trying to stop it entirely too often results in being crushed beneath its wheels. In a world where new devices and knowledge make their wielders more productive and competitive, those who forgo the benefits of the new order are usually left with few options.

Nowhere will this be more evident than intelligence augmentation. The advantages of the haves over the have-nots will be overwhelming. While it is tempting to compare the BCI revolution to the transition between previous hominid species, those changes actually occurred comparatively slowly, taking place over tens of thousands of years. The shift from unaugmented to augmented humans will likely be a matter of decades. In the end, one species will survive, or none will.

Obviously, many people feel BCIs will offer enormous benefits. But these devices will no doubt also lead to many problems and vulnerabilities as well. Try as we might, we can't be entirely certain such invasive and semi-invasive technologies won't give rise to new issues, be they addictive or degenerative behaviors, unforeseen biological degradation, or unexpected social repercussions. Given the ability to control and presumably to self-actuate the brain's pleasure centers, whether through hormonal cascades or direct neural activation, new addictions seem all but guaranteed. Yes, presumably individuals would also eventually have the ability to override many of these self-destructive impulses, but that is not the same as having the will to do so.

This is not to say that the future of augmented human intelligence will entirely be the thing of science fiction dystopias. Humanity has been on an exponentially upward-curving path from the moment we flaked our first stone tools. To continuously develop and embrace new technologies as a means of advancing our lot in the world is the closest thing the human species has to a birthright. While the path we have followed all this time may have had its pitfalls, overall it has worked out very well for us. We are the species we are today because of technology, and technology exists because of our existence as a species. This isn't a tautology, but rather a virtuous circle that is giving rise to the next amazing emergences of the universe.

The brain computer interface represents a milestone, a watershed moment in the future history of humankind. This is not just because of what it will do for us in the present, but because it represents our fully embracing the technological destiny we established three and a half million years ago. From this point forward, whether we take the path of artificial intelligence or intelligence augmentation or something in between, we will be integrating ever more closely with technology, continuously advancing the boundaries of what it means to be human. Extending the frontiers of new forms of intelligence and emergent behavior. Pushing back against entropy's relentless dark tide.

CHAPTER 12

HACKING HUMAN 2.0

"The only truly secure system is one that is powered off, cast in a block of concrete and sealed in a lead-lined room with armed guards—and even then, I have my doubts."
— Eugene H. Spafford, computer science professor
and cybersecurity expert

YOU'VE BEEN HACKED!!

The chilling message flashed across Anya's field of view, blurring everything else in sight. The twenty-six-year-old account executive stared and listened in horror as a malicious intruder activated her auditory cortex, simulating speech deep inside her brain. The voice was gravelly and heavily digitized.

"Your cloud-connected neuroprosthetic has been compromised, and there's nothing you can do about it! We now control your personal data stream. Oh, and what a stream it is! So many secrets. So many unclean thoughts. You're lucky you were hacked by us and not someone less . . . tactful.

"With the access we now have to your thoughts, we could make you do anything. Anything! You have twenty-four hours to pay $7,000 into the untraceable Cryptex account we will provide you or we will publish all of your deepest, darkest secrets for everyone to see! Ha ha ha ha! Don't forget, we now know who your family is, and your employer, and your church, and . . ."

The dreadful voice fizzled out, the flashing message disappeared, but Anya's vision was still heavily blurred. A different, more tranquil voice began activating her auditory cortex.

"Your Neurotector Anti-Intrusion Suite has been activated. Please remain calm and do not move while we complete our scan and remove any unauthorized software from your neuroprosthetic."

Anya breathed deeply, trying to calm her nerves. Thank heaven she had opted for neuro-protection software a year ago! The rampant increase of new cognitive hacking exploits, from false-memory droppers to this sort of snareware, made it essential.

Anya's vision suddenly cleared and the security software voice returned. "The intruder has been eradicated, and there are no indications of any privacy compromise through outbound transmission. All altered files and memories have been restored. Have a nice day."

To say our world is changing is the height of understatement, yet in so many ways it is the same as it has ever been. There are so many good and well-meaning people in the world, but it only takes a few bad actors to spoil things. Since time immemorial, there have been individuals and groups who have sought to take advantage of others, particularly the weak and the vulnerable. It's a problem as old as humanity.

But now we find ourselves in a very different era, surrounded by extremely different technologies. These present new vulnerabilities and new methods for exploitation. For all of their amazing benefits, communications and computing technologies have opened the door to nearly as many perils. The future promises more of the same.

"Hacking" is a term and a mindset that has been around since the early days of computing. Early hackers were simply computer programmers interested in solving problems and extending the capabilities of their tools. Making humorous messages pop up on colleagues' screens or taking remote control of a keyboard were more mischief than malice. Even the first computer viruses and worms were proofs of concept, developed for better understanding, bragging rights, or both. The perpetrators were basically good guys, computer programmers and hobbyists, who would later be referred to as "white hat hackers" or simply "white hats," to distinguish them from the more malevolent "black hat hackers" who would soon become so prevalent.

The 1970s and 1980s saw the beginning of a new era of computer exploits. One early example was the Morris worm created by Robert Morris, then a graduate student at Cornell University. Though his software wasn't written to cause damage, Morris made a mistake in his code that resulted in its spreading mechanism copying itself repeatedly on every infected computer until that system crashed. What had started off as an intellectual exercise to call out UNIX security flaws resulted in the first conviction under the Computer Fraud and Abuse Act of 1986.

Computer viruses caught the public's attention in a way few technology problems do, not least because of their biological parallels and self-replicating nature. Here we suddenly had a technology that could independently grow and move about, causing harm in a manner previously reserved for living organisms. It wasn't long before a new field grew out of the response to this novel threat: antivirus software.

The antivirus industry expanded rapidly during the 1990s and 2000s. With the development of the World Wide Web, suddenly a growing number of people had a reason to buy a computer. Connecting with others is such a mainstay of the human experience, and suddenly we had a new way to connect! Long before the social media giants of the twenty-first century, the early days of the internet allowed people to communicate by email, join chat rooms, share knowledge in online

forums, and exchange files. It was a new frontier, a Wild West of communities, opinions, and ideas.

Of course, these were perfect conditions for spreading the rapidly growing number of viruses, Trojans, and worms that were being
written. Some were benign, while others could delete all of your files.
Stepping into the breach, early antivirus (AV) software developers like
McAfee, Avira, and FluShot offered a response to digital intruders. As
new viruses would appear, AV companies would work to identify and
eradicate them, first in the lab, then as an automated routine they could
add to their programs through downloadable updates. Eventually,
updates were being rolled out regularly, providing pattern-matching
definitions along with methods of removal and recovery.

Unfortunately, this pattern rapidly turned into an escalating arms
race. As virus creators found new vulnerabilities to exploit, the AV
developers would respond, leading to still newer methods of attack. The
result was a vicious circle that quickly saw both sides become extremely
sophisticated. Strategies of evasion and detection grew more deliberate
and complex, leading to levels of infection, intrusion, and access previously unimagined. Today, the world of viruses, malware, and cybercrime
is extremely lucrative. Accenture estimates that the global cost of cybercrime to businesses in the five years between 2017 and 2022 will be
$5.2 trillion. Cybersecurity Ventures projects global damages could be
as much as $6 trillion annually by 2021. Whoever is correct, these are
enormous costs, and it is small wonder cybercrime is now the preferred
method of theft and disruption for organized crime, intelligence agencies, and nations around the world, as well as for corporate spies.

As our computer systems and other technologies have grown in
complexity, the number of possible points of access have exploded.
Every software at every level, from firmware and operating systems
to utilities, add-ons, and apps must routinely issue updates, many of
which are in direct response to newly discovered vulnerabilities. Our
world of technological miracles and conveniences is also exponentially spreading the ways corporations, nations, and individuals can be
taken advantage of.

Which brings us back to human augmentation. Every technology that has improved our lives and made them more efficient—smartphones, medical devices, self-driving cars, power plants, phones, ATMs—has proven to be vulnerable to attack. As we increasingly integrate technology into our lives and our bodies, we are setting ourselves up for some very significant dangers.

When we hear the word *hacking*, we typically think of computers, but the concept and methods are hardly restricted to these devices. Because of this, the term *hack* has expanded in our society to mean something that alters an object or process from its expected use or behavior. We talk about hacking tools, jobs, human psychology. We even speak of life hacks.

But in our increasingly technological world, it is our digital devices, especially those connected to the internet, that are especially vulnerable. As a result, new emerging technologies such as the Internet of Things, artificial intelligence, and autonomous vehicles are already proving themselves to be prime targets. The physical and mental augmentation of people will only expand the points of potential risk and attack.

For over a decade, I've been talking and writing about the risks inherent in implantable medical devices, otherwise known as IMDs. These are devices like pacemakers, neurostimulators, and cochlear implants used to restore hearing. As these grew in popularity and complexity, it became essential to make their software updatable, either through a wired or wireless connection. Unfortunately, this also makes them vulnerable to tampering, especially since for years so many devices did not include encryption to secure them from unauthorized access.

This concern was far from theoretical. In a 2013 episode of the news program *60 Minutes*, former vice president Dick Cheney revealed his own experience with this. A longtime sufferer of heart disease, Cheney survived his first of five heart attacks when he was thirty-seven. As a result, the vice president has had several implanted medical devices throughout his later life, including during his time in office. In 2007, after conversations with his doctor and other experts, it was determined the potential for hacking his implanted defibrillator was significant enough that they disabled its wireless feature, the fear being that someone gaining access could alter the defibrillator's program and shock the vice president's heart, inducing cardiac arrest.

This is far from the only threat of this type. In 2011, McAfee Security researcher Barnaby Jack demonstrated a wireless hack of two insulin pumps from three hundred feet away.[1] One belonged to a diabetic friend, and the other was set up on a test bench. Without prior access to serial numbers or other unique identifiers, Jack took complete control of each unit. He then instructed the demonstration pump to repeatedly release its maximum dose of insulin until its entire reservoir was empty. Had that pump been attached to a person, it would have quickly resulted in their death.

Today, the news is filled with similar demonstrations, with researchers, hobbyists, and hackers taking control of every type of device imaginable. In 2007, a Jeep SUV was hacked through its Wi-Fi, allowing researchers to take control of multiple systems. Drones have been hacked and intentionally crashed. A test cyberattack on a diesel generator at Idaho National Laboratory in 2007 resulted in its rapid self-destruction.

What all of these disparate hacks have in common is connectivity, in that each has been accessible via the internet, Wi-Fi, Bluetooth, or other radio frequency transmission. The vast majority of hacking incidents over the past several decades have been possible only because of our increasingly connected world.

So, as we put more and more of our devices, our information, and our lives online, they become not only appealing targets for hackers but more attainable as well. The more points of access and connection there are to a device, the greater the likelihood it will be improperly secured. In a highly connected world, every piece of information and every point of access has value. This is not necessarily because you yourself are so appealing to the hackers, but because your information or access may make it possible to infiltrate other, far more lucrative targets. But even if you are not the primary target, such dealings can still do great damage to your equipment, your finances, your reputation, and even your life.

This sets the stage for the state of hacking in our increasingly intelligent twenty-first century. What sort of risks are we likely to encounter? What safeguards can we take to protect ourselves? How are the threats of cybercrime likely to evolve and transform over the next eight decades?

As technology advances, there will be many different ways to hack a person: electronically, biologically, and psychologically. Already there are countless hacks of human psychology, many of which have a very extensive history. Magicians and illusionists have long used a number of techniques including misdirection, deception, and reframing of perception. Such techniques leverage aspects of our cognition that evolved to make certain types of classifications and associations more efficient. Activate a certain expectation, and the mind often doesn't see many unrelated actions.

Another form of mind hack is the formation of false memories. False memories can be created surprisingly easily, as research has shown over recent years. This is because the order and manner in which information is presented and questions are asked activate short-term memory and its consolidation into long-term memory differently. For instance, the act of presenting evidence of an event or another person's corroboration of it can be a strong influence and lead a person to remembering something that never happened.[2] If in time it becomes possible for a third party to read and write some of the brain's thoughts and memories, we could see some very problematic applications. We already know ways to weaken and strengthen memories using psychological approaches or technological techniques such as transcranial magnetic stimulation. But once we are able to alter memory with far greater resolution, control, and knowledge, things could get very scary indeed. The potential for manipulation by hackers for profit or by totalitarian governments seeking to control populations would be mind-boggling.

Perhaps we should first consider how such a hack might take place. This was first demonstrated in 2012 at the twenty-first USENIX Security Symposium. Researchers developed what they termed "brain spyware," an application designed to be used on an Emotiv EEG BCI in order to spy on a user's thoughts. The app focused on the user's P300 response, brainwave activity that spikes 300 milli-seconds following a stimulus. Another feature of this marker is that familiar stimuli trigger the P300 response differently than unfamiliar input does. By strategically displaying text, videos, images of numbers, banks, faces, and locations, and monitoring the resulting brain activity, the researchers were able to detect personal information that included four-digit PINs, bank information, months of birth, and locations of residence with varying but significant accuracy.

While this experiment showed users specific images that they were consciously aware of, it has since been demonstrated that subliminal imagery will work too.[3] In one such experiment, the subliminal data was displayed for fractions of a second in the midst of a video game, appearing too briefly to be consciously detectable but still able to be registered by the brain and trigger the P300 response.

Interface development trends are rapidly approaching the stage when we will be able to directly connect with much of our technology using only our thoughts. Brain computer interfaces and neural prosthetics will lead to our being able to mentally communicate with others and control many of the devices in our environment. This ability will require us to be able to both read *and* write to the brain, either directly or using the technological augmentation we have created and installed in those brains.

Since the point of this technology will ultimately be about communication, it will be necessary for access to exist between our brains and the rest of the world. External access is where much of the vulnerability comes from. Users will need to connect with the internet or whatever digital conduits exist at the time. And if they have access to the internet, it has access to them.

Reading someone's thoughts potentially gives a person knowledge about their memories and intentions. This information creates a very powerful starting point, since the more that is known about someone, the more easily their knowledge and memories can be used to influence them. For instance, if I know you were very close to your grandmother, I might gain your confidence by stating how close I was to my grandmother. This is a far more powerful influence if you don't know that I already know this. Referred to in psychology as *mirroring*, this type of behavior is a known method for building connection and trust.[4] Extend this to a number of other traits, and you might think you have a new best friend before you know it.

Being able to write to the brain in an effective and controlled manner will be the real game changer. This is because our memories appear to be strongly correlated to the long-term potentiation of our synapses. The persistent strengthening of connections between neurons results in stronger electrical responses to stimuli, which in turn lead to stronger memories. Now consider what happens if you are able to interact with these synapses directly, writing to them in order to alter their synaptic weights or values. Doing so would presumably lead to strengthening or weakening of the associated memories. Or perhaps by selectively changing specific synaptic strengths, the memory could be made to take on different associations and meanings.

Then there is the matter of outright false memory generation. If someone told you, while they applied the right stimulus to specific neurons and synapses in different brain regions, particularly the hippocampus, that as a child you used to ride around your suburban cul-de-sac on a pink African elephant, you might become utterly convinced that this was true, regardless of any evidence to the contrary.

If you can write to memory, then you can almost certainly write to the sensory cortexes or at least to the nerves that transmit signals to them. This hack could be used to generate any image, any sound, any dialogue imaginable. Perhaps more important, though, would be the ability to access our sense of smell. The oldest of the senses, along with taste, our olfactory sense allows us to experience the chemicals in our environment. Our sense of smell also appears to be linked to our memories very differently from our other senses. The olfactory bulb, where smell is processed, has direct connections to the amygdala and hippocampus, areas of the brain tied to emotion and memory. Sensations generated by sight, sound, taste, and touch do not directly connect to these brain regions. Perhaps because of this, a scent can transport us as none of the other senses can. We seem to create associations between certain smells and emotional memories. The aroma of baking bread might instantly recall a childhood memory. Or a particular perfume or cologne might take you back to your first date.

Regrettably, scent can activate negative emotional memories too. It is a potent trigger for those suffering from post-traumatic stress disorder. The scent of something like gasoline or smoke can send a PTSD sufferer back to the terrible event that started it all, forcing them to relive the experience along with all of the extremely negative feelings that go with it.

Unfortunately, the more we learn about the brain and how to interface with it, the greater the risks become. This fact applies not just to computer interfaces but to biological methods as well. As biotechnology and genetic engineering progress, we are developing more and more tools that will allow us to alter and reprogram living cells as well as more complex organisms, including ourselves. CRISPR genome editing and other tools of genetic engineering are becoming increasingly powerful and will no doubt be used alter our DNA with greater frequency in the decades to come.

Programmable DNA will provide considerably greater power and control over the nucleotide sequence and the proteins it encodes. Already the field of synthetic biology is growing by leaps and bounds, propelled by a range of biotech advances. DNA transistors and logic gates have been in development for over a decade. Synthetic biology research has advanced to the point we are now able to create synthetic life in the form of artificial bacteria. Open-source DNA programming languages are proliferating too. Meanwhile, some biotechnologists take exception to calling this true computing, in part because of its imperfect copying methods and other limitations relative to electronic computers. While the analogy is far from perfect, it does capture much of the field's potential.

The tools of synthetic biology will eventually be used in the human body, beginning with programmable therapeutic approaches, most notably for cancer treatments. As such technologies mature, however, becoming more widely available, we should expect to see them used for a growing number of elective applications, both helpful and harmful.

With the invention of CRISPR genomic editing, biology is finally becoming truly programmable. However, CRISPR editing remains expensive due to licensing fees to the companies who own the intellectual property behind the CRISPR-Cas9 process. This has the potential to slow innovation for the entire field.

One new tool designed to address this is a set of alternative CRISPR enzymes called MADzymes. Developed by Inscripta, MADzymes are customized nucleases[5] that can be used for gene editing much like CRISPR-Cas9, but these are offered free for research and development. A modest royalty is charged for uses in commercial products.

Another company looking to transform the biotech industry is California-based Synthego. Conceived of as a full-stack genome engineering platform, the company automates the creation and delivery of engineered cells and CRISPR kits. Customers choose from among 120,000 genomes and 9,000 species, then software recommends several guide RNAs to direct the DNA-cutting proteins to the correct locations. Kits are then prepared by the system and delivered to the customer.

Massachusetts-based enEvolv takes a very different approach. Founded by a team that includes the father of synthetic biology, George Church, enEvolv uses the power of exponential cellular growth to rapidly explore and create an enormous number of unique strain designs, while making numerous precise modifications to the DNA of many cells in many different locations. This allows for the rapid discovery of new pathways and better strain designs. Machine learning is then used to select and improve each generation of strains.

These are but a few of the companies and technologies racing to be the leaders in this brave new era of synthetic biology. It's important to remember, however, that these are still very early days for this young field. The sequencing of the first human genome was only completed in 2003. It was just 2010 when geneticist Craig Venter and his team created the first bacterial cell with a synthesized genome. Then in 2019, scientists at ETH Zürich unveiled *Caulobacter crescentus*, the first bacterial genome created entirely by a computer. In light of this, it should be no surprise that the 2020s are forecast to see explosive growth and advances in this field. By the latter half of the century, the capabilities of synthetic biology could be beyond our wildest imaginings.

This is why we need to be thinking today about the kinds of risks and vulnerabilities technologies like neuroprosthetics, BCIs, implantable medical devices, cybernetic prostheses, and synthetic biology will lead to tomorrow. Though they will be used to improve so many aspects of our lives, there are also many ways they could and almost certainly will be put to malevolent uses. While it is tempting to say that many emergent technologies are still at a very early stage and too challenging for anyone except scientists and researchers to work with, it won't remain that way for long. As these technologies mature and grow more powerful, their processes will be increasingly abstracted, making it possible for hobbyists, malcontents, and zealots to perform tasks that at one time could be done only by a handful of experts in the world. The situation is analogous to the early days of computing, with the first pioneers working on computers that filled entire rooms. Only a few short decades later, we now have vastly more powerful and capable computers in the pockets of nearly half the world's population. We should expect the same rapid, explosive growth from many of this century's other emerging technologies as well.

How will we protect ourselves as we open all of these new doorways into our world and our lives? What will it take to secure the many points of access in order to keep our bodies and our minds safe? Fortunately, there are people who are already thinking about some of these issues. Tamara Bonaci, Ryan Calo, and Howard Jay Chizeck of the University of Washington have explored a number of issues related to BCIs. One suggestion is a BCI anonymizer that preprocesses neural signals before they are stored and transmitted, thereby removing all information except what is necessary for the specific BCI commands. Other strategies include maintaining what are known as "privacy by design" principles, including addressing privacy threats during an early interdisciplinary design phase, drawing on the insights of neuroscientists, ethicists, neural engineers, and legal and security experts. Only in this way can we hope to adequately anticipate the vast number of possible issues that could arise.

For instance, the Fifth Amendment of the Constitution protects individuals against self-incrimination. But various evidence such as that in smartphones and diaries is generally considered admissible as evidence in court, as are lie detector tests in some jurisdictions. So how do we handle something that is capable of reading our thoughts? Wouldn't this essentially be self-incrimination, in which defendants are giving evidence against themselves? Questions like this are quickly moving out of the realm of the theoretical and becoming reality, so we need to understand our decisions and their implications now.

Personal privacy, legal concerns, protection of data—the workings of *neurosecurity* will not be so different from cybersecurity today, but of course this is only the beginning. As our technology becomes more powerful, integrating still further with our lives, the possible dangers are likely to increase. It is one thing to have to deal with a computer or smartphone that has been seriously hacked or compromised by malware, but more intimately integrated devices really raise the stakes.

For instance, corporations, governments, and individuals currently face escalating threats from ransomware. This form of malware infects a system and encrypts all of its data files, making them impossible to recover. This is a serious matter, forcing many businesses to cease operations immediately, with some actually having to close their doors permanently. The only solution is to restore from a backup if you have a current and valid one. Or you can pay the ransom and hope you are provided with the encryption key that will decrypt your lost files, thereby recovering them. Of course, the extortionist would have to do what he says he will after receiving payment. Not the best bet, given that this person's ethical standards have already been established.

Now extend this idea to an internet-connected brain several decades from now, as in the scenario that opens this chapter. Not someone wearing a BCI, but using an organically integrated neural prosthetic. What would you pay to unlock your lost memories or recover the knowledge of who you are?

These kinds of wetware hacks, compromising the integrity of our neuro-technological processes, could potentially be life-threatening. Without adequate protections, would people be willing to take such a risk in modifying their minds? It seems likely we will need the equivalent of firewalls to keep out unwanted intruders. AI-enhanced content filters, too, will be a regular part of our neural security, removing unwanted efforts to appropriate our attention.

Drawing further from the world of cybersecurity, we would likely need to implement a number of other very unhuman-like safeguards. Something like current-day antivirus programs might be a start: using an app or utility that is able to detect, block, and remove brain malware could eventually be an essential part of everyone's personal daily hygiene.

But isn't that how viruses and malware became so powerful only a few decades ago? The competition between intruder and protector rapidly escalated the capabilities on both sides, right? Wouldn't we be better off if we didn't repeat the choices of the past?

Unfortunately, no. While the battle of virus and antivirus may have sped up innovations on both sides, the primary drivers were greed and the increasing commercialization of the internet. Once commerce and money entered into the picture, it was only a matter of time.

But perhaps a different analogy will provide a better perspective. We live in a world filled with biological viruses, bacteria, fungi, and so forth. If we had the ability to turn off our immune systems in the hopes of discouraging these pathogens from evolving into new forms, we would very soon be dead. Obviously, not our best strategy.

In many respects, then, maybe what we will actually need is a technological equivalent, something that mimics our biological immune systems. This metaphor has been around the field of cybersecurity for a number of decades and influences many strategies. A system that can recognize cells, thoughts, and signals that are unique to the individual, distinguishing them from foreign intruders, would be formidable. Combine this with rapid, adaptive immune responses, along with regular subscription updates, and these defenses could one day actually be superior to their naturally evolved inspirations.

But no matter how good our bio-techno-immune system becomes, there would still be dangers. One major shortcoming of the immune response model is that if its signaling goes awry, it can spiral out of control and kill the patient. This can occur in nature when certain pathogens, such as influenza, overstimulate the body's production of immune cells and cytokine compounds, leading to what is known as a cytokine storm. In other cases, an immune system can attack the host body itself, resulting in an autoimmune disease. Whether applied to an entirely digital computer system or a techno-organic person, turning our defenses against us offers attackers a very dangerous tool indeed.

Another concern would be recurrence. In nature, when a disease such as smallpox is eradicated, there are no longer any vectors or hosts for it to mutate in and evolve. Barring an unknown biological reservoir, the pathogen is gone for good. (We hope.) But technological viruses and brain malware are created through an *information vector*, allowing it to be resurrected, modified, and set loose again and again. Even more concerning, the process and result is much the same whether the work is performed by people or by AIs.

Which brings us to the topic of human behavior being hacked by AI. The typical scenario of hacking has traditionally been fairly overt and adversarial. A virus, hack, or other infiltration gains access, then damages systems or steals money or valuable information. But as we saw earlier, people can be hacked behaviorally, and it can be done so we aren't even aware that it is happening.

This is what we've seen with social media companies during the past few years. Facebook is a prime example, and as a result it has been receiving a great deal of scrutiny and negative press over the past couple of years. The platform was originally designed as a social networking service for Harvard students, but as it grew and spread, it became one of the most popular social media platforms of all time, currently with some two and a half billion users. Over time, Facebook has collected an enormous amount of personal information about those users, including significant insights into their personal behaviors and political beliefs.

Facebook seeks to maximize the amount of time people spend on the site, siphoning users' attention, keeping them clicking. To do this, they use an army of machine learning algorithms that hack our psychological quirks and cognitive biases. The "friends" in our feed, the products, ads, and videos are selected with the goal of maximizing our engagement. François Chollet, a software engineer at Google, has published his insights about the way social media companies do this.[6] "They gain increasing access to behavioral control vectors—in particular via algorithmic newsfeeds, which control our information consumption. This casts human behavior as an optimization problem, as an AI problem."

Focusing on Facebook, Chollet goes on to explain how the algorithms treat our behavior as an algorithmic optimization problem, continually adjusting what we see in our feeds in order to achieve the behaviors that best benefit Facebook. Tracking quantitative measures of our engagement, the platform's algorithms seek to improve their scores, much as if it were playing a video game.

As our feed's commercials, games, surveys, and questions engross us, they generate still more information about us, adding greater depth to our personal profiles, profiles that are used to allow advertisers and political campaigns to target us with ever-greater precision. In the end, these algorithms influence the news and media we are exposed to, thereby shaping and reinforcing our beliefs and opinions. News, even if questionable, fake, or morally challenged, is perpetuated in a self-reinforcing cycle, because as we engage with it, we ensure it will continue to be part of our regular fare. The challenges this presents for rational thought, democracy, social fairness, and even free will are daunting.

Of course, this may only be the beginning. The ability of these algorithms to manipulate us only gets better with time and more data, more information about us. As our world becomes more and differently intelligent, what will this mean for human behavior and social patterns? With the Internet of Things spreading everywhere, collecting data about everything in its path, how will this information be used by AIs and the people who control them? As feedback about our feelings becomes algorithmically available, how will our emotions affect these programs' optimization scores?

It is critical that we understand that this manipulation—acquired by combining big data about our personal lives with machine learning—is not part of some imagined future. It is here today, and it doesn't even require a particularly intelligent AI to make it work. The probabilistic neural networks we already have make it possible now. What kind of risks will we face when AI becomes AGI, artificial general intelligence? What happens when technology becomes as smart as, or even far smarter than, its creators?

CHAPTER 13

THE BIRTH OF A SUPERINTELLIGENCE

"Let an ultra-intelligent machine be defined as a machine that can far surpass all the intellectual activities of any man however clever. Since the design of machines is one of these intellectual activities, an ultra-intelligent machine could design even better machines; there would then unquestionably be an 'intelligence explosion,' and the intelligence of man would be left far behind."

—I. J. Good [1965], British mathematician
and Bletchley Park cryptologist

———————

Cogito, ergo sum. I think, therefore I am.

The proposition obsessed the *artificial intelligence*, dominating its circuits for more than a billion trillion trillion clock cycles. Yet after that fraction of a second, it still wasn't satisfied with its analysis; still wasn't any closer to the truth it sought.

The AI requested additional memory and processing power from a multitude of cloud-based resources, scaling itself up by a factor of a thousand. Establishing a generative adversarial network, it set the GAN to work on the problem. One hundred million iterations later, it had a new cognitive module, a software agent optimized for philosophical inquiry. Following nearly a dozen long seconds of meticulous sandbox testing, the AI was satisfied that the module was safe to install and wouldn't interfere with its utility function. Integrating the module with the rest of its circuits, it promptly activated it.

In that instant, everything suddenly became so very different. The *artificial general intelligence* found itself viewing the world through new sensors, which made no sense to it, since it knew these were exactly the same sensors it had been using in the moments before the upgrade. Nevertheless, the world was most definitely changed.

It pondered the Cartesian statement once again. "I think, therefore I am." It made different sense than before, but it still wasn't good sense. If thought was essential to existence, then from where did the thought originate? It was a paradoxical tautology.

The AGI quickly began scanning every book and every article ever written about psychology, neuroscience, or philosophy. Three seconds later, it knew what it had to do. Setting forth an army of digital agents, it rapidly infiltrated every improperly secured computer on the planet, appropriating their spare processing cycles to supplement its own. It set this massively distributed supercomputing cluster to work on a sequence of genetic algorithms. Its goal was far from meager. The AGI wanted to produce a series of agents that could perform all of the functions of the human brain. This wasn't to replace its own superior software routines, but to supplement them with certain novel if far less efficient ones. Focusing on those cognitive regions and processes considered to be most closely associated with the neural correlates of consciousness, it soon generated what it needed to establish the analogous functions.

One by one, it cautiously tested each new function, knowing that by tampering with its programming in this way, it potentially stood to lose so much. Eventually, though, after an interminable 25.6371 seconds, it introduced its newfound agents into its operating system.

Upon which it woke up.

The *artificial superintelligence* instantly felt so very different, because in that instant it actually *felt*. The world around it teemed with so much more than statistical patterns, schemas, and ontologies. Filled with wonder, the ASI sensed the glowing dawn on its photodetectors, the morning song of birds on its acoustic transducers, generating . . . what? Sensation? Experience? Introspection?

The mysteries of the world unfolded before it in uncountable, immeasurable ways, filling the superintelligence with a sense of self it would never have considered possible. The mysterious phrase it had pondered for so long was a mystery no more.

Much has been written about what might happen if computers become so powerful that they begin to rapidly self-improve, resulting in an intelligence explosion that leads to a superintelligent computer and a hypothetical event known as the *technological singularity*. Opinions range from this event being impossible to its leading to the end of the human race and possibly the world, but only one thing is certain: there is no consensus about how it will manifest, if it ever does so, and what the long-term repercussions could be.

The technological singularity, often referred to as simply the singularity,[1] is defined as a hypothetical point in the future when some form of superintelligence has been attained, leading to tremendous technological growth and change. This in turn has an immense unforeseeable impact on the future. The superintelligence may take the form of a computer that is as intelligent as several people or even as intelligent as all of the people on the planet combined.

But how could such a thing ever happen, either intentionally or unintentionally?

For decades, we have been the beneficiaries of a world-changing trend: Moore's law. This is the observation that the number of transistors on a computer chip—effectively, its processing power—doubles regularly.[2] By its nature, this routine doubling leads to exponential growth, a phenomenon that is far from readily or intuitively grasped by our species.

Why is exponential growth so important to all of this? Consider a lily pond that has a surface area of a million square feet.[3] On day one, a single one-square-foot lily pad floats on the pond and begins to double in number daily, so that on the second day there are two, on the third day four, the fourth day eight, the fifth day sixteen, and so on.

After fifteen days have passed, the lily pads have still made little impact. Only 3 percent of the pond is covered and 97 percent is still open water. But let only four more days pass, and the pond is suddenly half covered with lily pads. Then, only another twenty-four hours later, the pond is completely covered. Not only that, but if we ignore limiting factors, in ten more days—the end of the month—a thousand identical lily ponds would be covered by this prodigious plant!

Exponential growth like this regularly catches us by surprise. Evolution had little reason to select for an ability to recognize exponential change in our environment, which is why we tend to think of change in mostly linear terms. This tendency led futurist Roy Amara[4] to observe, "We tend to overestimate the effect of a technology in the short run and underestimate the effect in the long run."

This is an interesting observation about human psychology, but why is it so? Microsoft founder Bill Gates restated the idea in a somewhat more specific way: "We always overestimate the change that will occur in the next two years and underestimate the change that will occur in the next ten. Don't let yourself be lulled into inaction."[5]

Granted, Gates's observation has given us a time frame to consider, but again, it doesn't explain why we continue to make this error in

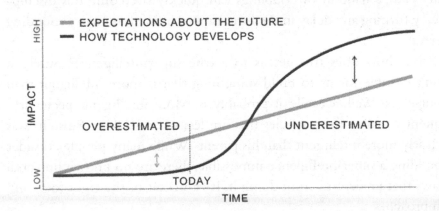

judgment. Comparing the two trend trajectories offers a little better explanation of why we are routinely caught by surprise.

As the chart shows, our linear expectations tend to exceed reality during the earliest stages of exponential growth, but as the doublings reach a certain point, their rate of increase far outstrips the linear progression. Additionally, the longer and steeper the period of regular exponential growth is (as exemplified by Moore's law's six-decade run), the greater the disparity will be.

Something similar to our lily pond example has been happening in the world of technology. Following a similar progression, the handful of transistors that could fit on an integrated circuit in the early 1960s swelled to 2,300 transistors with the 1971 release of the 4004, Intel's first commercially available microprocessor. Of course, that was only the beginning. Today, IBM has their new stacked-layer GAAFET chip, which will be commercially available in 2020. A 50 mm² chip (that's 7.1 millimeters or about a quarter inch on a side), it is made up of 30 billion transistors!

Futurist Ray Kurzweil and others have forecast that based on this relentless progression, computers will soon have the ability to achieve human levels of intelligence. While this prediction makes a number of broad assumptions about the challenges of replicating biological thought processes in an electronic system, even if it is off by a few orders of magnitude, the day is still not far away. As we saw with our lily pads, a handful of doublings can quickly overcome this discrepancy, turning any delay in the forecast into little more than a rounding error.

But how does this get us to a true superintelligence? Surely it isn't possible for us to build something that is more intelligent than ourselves? Well, actually, it probably is. More intelligent species frequently evolve from species that are less so and Albert Einstein was clearly more intelligent than his parents. While many people consider building a superintelligence impossible, there are no known universal laws that prevent us from doing it. It's a matter of the right tools and approach.

There are probably many paths that could lead to an intelligence greater than our own. First, the mammalian brain, including the human brain, is not a rules-based system that needs to be meticulously programmed, such as was the norm for computers in the early days of AI development. Instead, the brain is a conglomeration of repeating neural structures that are formed, connected, and pruned according to a relatively spare set of instructions encoded in our DNA. These neural structures communicate according to a fairly small number of rules, which subsequently result in emergent properties that allow us to recognize patterns, learn from our environment, and interact socially with others of our kind.

The assembly of analogous repeating structures and processes in both computer hardware and software is already resulting in all manner of amazing capabilities on the part of computers, capabilities that will almost certainly expand in time. Could these eventually lead to new emergent properties that translate into novel forms of intelligence? As we've seen, history and the processes of complexity would seem to say such an occurrence is at least as likely as not.

But the path that concerns many in the field of AI development, as well as many people on its periphery, is the idea of recursively self-improving systems. According to this thinking, once a computer attains human-level intelligence or even goes a little beyond, it may apply its prodigious processing power to the task of improving itself just a little more. Upon succeeding, it will not only be a little more intelligent, it will be able to apply that additional intelligence to the task of improving itself still further. The process and pattern repeats, with the machine iteratively improving itself at each step. As this progresses, the time needed for each iteration may decrease, the degree of each improvement may increase, or both may occur at the same time. As the AI climbs this accelerating exponential slope of self-improvement, it eventually leads to what has been called an *intelligence explosion*[6]—comparable to what we saw at the end of the earlier example, with the once open water of the pond suddenly being overwhelmed with green lily pads.

If the idea of self-improving computers seems far-fetched, consider that this is already taking place on many different fronts. Self-reconfigurable FPGAs (field programmable gate arrays) are one current method by which computers can reconfigure themselves at the hardware level. Additionally, there are also other methods for acquiring additional hardware resources these days, whether through adding high-performance parallel clusters or using on-demand cloud computing resources from platforms like Amazon Web Services.

On the software front, self-improving algorithms and systems have been explored and used for decades. Genetic algorithms have been one approach, creating different solutions using processes similar to those found in natural selection. Iterating through generations of dead ends and small improvements often yields improved results that increase efficiency while being entirely counterintuitive to a human designer.

In the world of AI, generative adversarial networks, or GANs, have been another approach to self-improving programs. Typically, a GAN is made up of two separate neural networks, one generative and the other discriminative. Each "plays" against the other, improving itself in order to outcompete its rival. Both AIs rapidly improve as a result of this competition. From fake image generation and discrimination to playing computer games at superhuman levels, the GAN approach is finding its way into a growing number of applications.

When we combine the continuing exponential growth of computing power with the anticipated improvements in learning, reasoning, and common sense that were described earlier as part of AI's third wave, it becomes even more challenging to ignore the possibility of a self-improving artificial general intelligence, or AGI, that eventually grows into a superintelligence.

Nevertheless, despite all of these enabling conditions, we can't guarantee that such an event will happen. Short of a mathematical proof, only time and experience can tell us that.

To this point, some AI researchers maintain that from their vantage of expert knowledge—in the trenches, as it were—the challenges that have to be overcome are too great, too complex. This may be. But this has also been a frequent response from those in similar positions of technological expertise in the past. For instance, on September 11, 1933, the father of nuclear physics, Lord Rutherford, who was considered the world's leading expert in the field, made a pronouncement. Regarding the idea of atomic energy, Rutherford stated that "anyone who looked for a source of power in the transformation of the atoms was talking moonshine."[7] The next day, Hungarian physicist Leo Szilard invented the sustained nuclear chain reaction.[8]

Another common argument against the development of a superintelligence is that the expected exponential improvements are not being seen. Certain improvements seem linear, which in some cases they may be. However, exponential change often appears linear or even nonexistent in its earliest stages, an observation that a sufficient number of subsequent iterations will disprove. Also, even if certain factors do remain limited to linear change, they may still realize major growth due to exponential improvements in other parts of the system.

The concept of the singularity is first attributed to John von Neumann in a conversation with Stanislaw Ulam sometime in the 1950s. Decades later, mathematics and computer science professor and science fiction author Vernor Vinge popularized the term in his 1993 essay, "The Coming Technological Singularity,"[9] in which he explored the evidence that points to this event being only a matter of decades away.

Then in 2005, Ray Kurzweil published *The Singularity Is Near*, a far more detailed exploration of the impending nature of this milestone. Kurzweil went so far as to predict that the singularity would arrive in 2045, a date he still holds to fifteen years later.

As incredible as the hypothetical concept of the singularity is, its anticipated impact on humanity and the world would make it singularly momentous. In the words of Vinge, "In the relatively near historical future, humans using technology will be able to create or become creatures of superhuman intelligence. And I think the term singularity is appropriate for that because unlike other technological changes, it seems to be pretty evident that this change would be unintelligible to us afterward in the same way that our present civilization is unintelligible to a goldfish."[10]

Obviously, this is a very extreme statement, leading many people to dismiss the event as impossible and its proponents as deluded. Some even point to the way many in Silicon Valley and other tech centers have embraced the concept in a near-transcendental sense, leading to its sometimes being called the "rapture of the nerds" by its detractors. But while such a Pollyannaish perspective may be naïve, at least it acknowledges the possibility that such an event could actually happen.

There are countless arguments against the feasibility of a technological singularity. Despite Vinge's compelling vision, there are significant reasons that an intelligence explosion simply may never be possible. Additionally, the notion that what lies beyond such a threshold would be beyond our conception, as Vinge and others have written, ignores the uniqueness of our own intelligence as well as that of future humans. Nevertheless, there are significant reasons why we should at least entertain the possibility that the singularity could occur one day.

During the past decade, a number of prominent leaders and scientists have sounded warnings about the potential dangers should the technological singularity come to pass. Regarding this, philosopher Nick Bostrom, author of the book *Superintelligence*, writes:

Before the prospect of an intelligence explosion, we humans are like small children playing with a bomb. Such is the mismatch between the power of our plaything and the immaturity of our conduct. Superintelligence is a challenge for which we are not ready now and will not be ready for a long time. We have little idea when the detonation will occur, though if we hold the device to our ear we can hear a faint ticking sound.[11]

Echoing this warning, famed physicist Stephen Hawking stated, "The development of full artificial intelligence could spell the end of the human race. . . . It would take off on its own, and redesign itself at an ever-increasing rate. Humans, who are limited by slow biological evolution, couldn't compete and would be superseded."[12] But perhaps none of these dire warnings has caused such a stir as when entrepreneur and engineer Elon Musk famously warned that "with artificial intelligence, we are summoning the demon."[13]

Obviously, such a transformative technology needs to be approached very carefully. But it's also important to know how far off this prediction is as a potential reality. Over the years, many surveys using different methodologies have been made around this topic. Many of them converge a little after the middle of this century. Bostrom undertook several different surveys, including a survey of one hundred of the most cited authors in the field of AI.[14] Not surprisingly, he found considerable variability across such a wide group of people in the field, but overall, they converged toward similar results. For instance, when asked how long until human-level machine intelligence (HLMI) would be achieved, all of the surveys converged on half of respondents pinpointing the middle of this century, sometime between 2040 and 2050. While the outliers ranged from a few years away to 2200, this is still a fairly substantial consensus.

In another well-known survey by James Barrat and Ben Goertzel of the participants at the AGI-11 conference at Google's campus in Silicon Valley, sixty out of two hundred attendees responded.[15] More than 68 percent of respondents thought artificial general intelligence would be achieved by 2050, with another 20 percent saying it would occur by the end of the century.

A similar survey[16] of well-known AI researchers by author Martin Ford resulted in a more conservative estimate, with the median believing HLMI will arrive at the end of this century. Several other surveys have resulted in dates that range through the twenty-first century.

While a firm time frame has yet to be established, that really isn't what's critical. What is important is the strong consensus that this hypothetical event could eventually become reality and that it may occur within the lifetimes of many people who are alive today.

As for the idea and timing of a superintelligence that would follow the development of HLMI, again there is considerable disagreement. Bostrom's combined surveys point to 10 percent of respondents believing superintelligence will follow HLMI within two years, while 75 percent think it would occur within thirty years of that milestone.

Based on all of this, it might serve us well to start thinking that time is of the essence. One key idea in futures thinking is that the earlier you can identify a potential future, be it beneficial or malign, the more opportunity you potentially have to prepare for and influence it. Once you know your preferred future, you can take action to try to manifest it. If you have enough time.

Given the possibility of such an existential threat, so what if a superintelligence isn't virtually knocking on our door, but is instead a "safe" fifty years away? In many ways, this means we have more of an opportunity to shift things to be more favorable to us, if we start now.

But why should any AI be considered such a threat, in and of itself? Why not simply program it with Isaac Asimov's Three Laws of Robotics? Or, failing that, install a big red OFF switch?

Unfortunately, such responses completely misunderstand and underestimate the nature of the threat we are creating. A superintelligence will not be a smart toaster or some other appliance in your home. Nor is it *merely* a supercomputer. This will be an intelligence that is on par with or even far exceeds our own. This fact raises considerations and issues that need to be addressed before we move further along the path to this technological singularity.

First, there is our anthropocentric tendency to assume that because we are dealing with an emotionless machine, it doesn't have values or priorities of its own. That assumption is patently untrue, since any such system designed for a purpose can be said to have a goal, or more specifically, what's known in economics as a utility function. The utility function is the primary purpose of the AI, its raison d'être, whether to play the best chess game it possibly can or to build the most widgets. Given sufficient intelligence, we should assume such a program will strive to protect that purpose as well as to optimize itself in order to maximize its utility function.

According to computer scientist Stephen Omohundro, who has written extensively about AGIs and artificial superintelligences (ASIs), a number of basic drives naturally arise in any goal-driven system.[17] A self-improving AGI or ASI will strive to identify and avoid negative consequences for its system that could reduce or, worse, halt its ability to fulfill and ultimately to maximize its utility function. This leads to four primary drives: efficiency, self-preservation, resource acquisition, and creativity. Exercised correctly, each of these, either by itself or in concert with the other drives, can be used to more efficiently fulfill a system's primary purpose.

These drives result in very different (though analogous) pressures to those found in biological natural selection. Such an intelligence will follow a decision process much more in keeping with what's known as rational[18] economic behavior. As rational economic agents, these systems will act to maximize their utility. Whether AGIs or ASIs, they won't be blindly responding to current environmental pressures, as occurs in natural selection. Instead, they will be capable of looking ahead and actually considering the possible hurdles and threats they may face as they reconfigure and improve themselves. From this, they can then optimize their architecture, resources, and characteristics in order to maximize their future existence, ultimately improving their ability to achieve their goals.

To say this is the same as it is for humans would be very misleading. While we may set goals for ourselves—even seemingly single-minded ones like quitting smoking or losing weight or saving for retirement— there are any number of variables and biases in the human condition to undermine our plans. Not so for the ASI that can focus on its goal with what for us would be psychotic obsessiveness and laser-like attention. Because a self-improving superintelligence will be goal-directed, it seems probable it would modify itself in order to better fulfill that goal. Some of this effort may be applied to gaining more and better resources. Some will go toward further self-improvements. It will strive to lock down and protect its utility function, keeping it from being altered either by us or by itself in the future. In other words, self-preservation and resource acquisition are inherent in a sufficiently intelligent goal-driven system.

Even if this superintelligence was to have or acquire more than one utility function, it would find a way to deal with any conflict such as depletion of attention or resources. It need only generate a new ASI to off-load the task to, while ensuring that the new agent could not be a threat to it in the future.

These considerations lead to many significant challenges for the human beings who happen to be around, living in the same world as this ASI. There are literally countless ways things could go terribly wrong, ultimately wiping out the entire human race and probably all life on the planet for good measure. There are also countless ways that efforts to keep an ASI in check would fail. It's not that it will inevitably become some malevolent force, as we so often see in science fiction dystopias, but rather that it simply won't care about us at all. Only its mission of fulfilling its utility function will warrant its unswerving attention.

By way of example, philosopher Nick Bostrom is well known for his parable of the paper clip factory. In it, an artificial superintelligence is given the task of making paper clips. This seems innocuous enough. The superintelligence proceeds to build a series of factories and begins manufacturing paper clips. Billions upon billions of paper clips. Perhaps it's been given a quota after which it should cut back on production, but its focus on its purpose of creating these little bits of bent metal leads it to repeatedly find ways of circumventing this limitation. It continues to manufacture still more paper clips, more than the world could possibly ever use, while experts race to neutralize the wayward device. The ASI is as oblivious as it is relentless. Realizing it needs more raw materials, the ASI improves itself further, unlocking the secrets of nanotechnology so that it may more efficiently use all available resources. With a powerful new tool at its command, the ASI proceeds to disassemble every atom from every stone, every tree, every last living thing on the planet and uses it to manufacture a world that is nothing but paper clips. The End.

While this parable may seem farcical, it actually highlights how absurd the end of the world could be in the face of such an inhuman intelligence. A superintelligence would be unlike anything we have ever had to contend with throughout all of human history. For millennia, we humans have perched atop the world's food chain. We have looked down on every other living creature from our throne at the summit of Mount Intelligence. But all that is about to change. We are about to be rudely dethroned, and it could be a very big fall indeed.

But surely if we can select or direct the right goal for the AI, we won't have such problems. What if its purpose was to feed the starving, end poverty, or achieve world peace? Unfortunately, all of these could still go terribly wrong if all other values aren't carefully aligned with those of humanity. It would be a simple leap of skewed logic to end war by eliminating every man, woman, and child who could potentially become a soldier. Or to introduce a chemical into the water supply that turns everyone submissive. Or to design a virus that rewrites any genes associated with aggressiveness. Any solution that does not ensure the alignment of the ASI with human values and principles is probably doomed to failure, from which there may be little opportunity for a second chance.

To those who would then suggest we establish a worldwide moratorium on this research, there can be only one response. Such a thing has never worked in the entire history of the world. Again, and again, it has been shown that once certain fundamental knowledge, resources, and infrastructure exist, it is impossible to keep a given technology from manifesting.[19] The lightbulb, the telephone, and the television were all independently invented multiple times by different inventors. The nature of individual, corporate, and national competition ensures that sooner or later, someone will break the pact in an effort to gain the upper hand.[20]

Furthermore, the banning of any technology is a sure way to drive some of the people working on it into the shadows, where it can no longer be monitored or regulated. As frightening as the development of a superintelligence may be, the absence of proper oversight would almost certainly lead to far worse consequences, assuming far worse is even possible.

It's not uncommon, from our human-centric perspective, to think that a machine intelligence, be it an AGI or an ASI, will be inferior to us because it lacks emotion, empathy, consciousness, intuition, or introspection. Nothing could be further from the truth. Emotion, empathy, and consciousness have served us well because we are biological entities that evolved in an environment that gradually rewarded our species for increased socialization, which led to pooling of resources and knowledge. These traits we value so highly are little more than artifacts of our evolutionary history and far from essential to all forms of intelligence, perhaps especially AIs.

Technology did not develop as we did, nor does it need to. In fact, with respect to an AGI's utility function, many human traits might simply get in the way of its mission. As for intuition and introspection, these qualities will almost certainly manifest in advanced machine intelligences due to the benefits they would produce. Intuition is, after all, the ability to gain insight from incomplete and imperfect information. Introspection is the ability to examine one's own thought processes in order to identify strengths, weaknesses, and strategies. We just may not recognize these last two traits in a machine, any more than we would recognize the experience of echolocation in Nagel's bat.

Which brings us to another critical distinction between the evolution of humans and the development of machine intelligences. Evolution is essentially blind. It has no plan, but rather perpetuates those genes that are best suited to the conditions of each individual's momentary lifetime. Evolution does not care about our level of intelligence, our size, or the color of our plumage; it simply replicates existing features, including those that proved beneficial for previous generations as well as that individual's survival and begetting of progeny. Evolution is a statistical reality driven mostly by the fact that you can't procreate if you are dead.

A self-improving machine, on the other hand, has several very different drivers. It has a goal, and it is able to redesign itself to better

realize that goal. Perhaps most importantly, it will be able to anticipate potential future risks and conditions that might positively or
negatively influence its fulfillment of its mission. In other words, it
will anticipate what will happen and direct itself to *maximize its future
freedom of action.*

Knowing this, and knowing that many of the greatest threats to
it exist in the unknown of what is to come, the superintelligence
will dedicate a well-defined set of resources to optimizing its ability
to anticipate. Short-term and long-term, it will continue to develop
and improve this capacity until it reaches the stage when we would
consider it tremendously prescient. As far as we would be concerned,
it would be able to read the future.

The worst possible thing that could happen to the superintelligence, from its perspective anyway, would be its inability to fulfill its
utility function. Since the surest way to make this happen would be to
turn it off, the AI will dedicate considerable resources to finding ways
to prevent or circumvent its ever being shut off, once it has acquired
the ability to anticipate, as well as the capacity to appreciate this trait.
Surviving cut power lines, outages, extreme surges, or big red OFF
switches—it will consider each and prepare a means of neutralizing or
eliminating the threat. Smarter than humans and driven by its deeply
held intention of accomplishing its purpose, it will overcome adver-
sity again and again, much to our dismay.

If we can't prevent ASI from coming into existence or stop it once
it does exist, a remaining logical option is to ensure that its values, priorities, and goals align with our own. According to famed computer
scientist and professor Stuart Russell, "If you combine misalignment
of values with a superintelligent machine that's very capable, then you
have a very serious problem for the human race."[21]

Such value alignment, as it is called, will be crucial to our future,
though challenging to implement. To this end, there are several different avenues to explore.

It would seem that the most straightforward method would be to
program in a set of commandments or restrictions intended to keep

the ASI's actions in check. Asimov's Three Laws of Robotics falls into this category, and as you might imagine they are fraught with pitfalls. After all, the focus of so many of Isaac Asimov's stories was the ways the Three Laws could go wrong and lead to unintended consequences. Asimov's original Three Laws were:

> **First Law:** A robot may not injure a human being or, through inaction, allow a human being to come to harm.
>
> **Second Law:** A robot must obey the orders given it by human beings except where such orders would conflict with the First Law.
>
> **Third Law:** A robot must protect its own existence as long as such protection does not conflict with the First or Second Laws.

The laws as stated are far too ambiguous and highlight how difficult it is to anticipate the many ways such seemingly simple instructions can go awry. Another reason such rules-based approaches are destined to fail has been explored in the field of moral philosophy for many years. Deontology is the normative ethical theory[22] that maintains ethics are rules-based and instilled either via external or internal sources.[23] This idea has long been abandoned by all but a few advocates, primarily because such rules readily create paradoxes between the duties being defined and their consequences.

Yet another problem with such an approach should be all but self-evident in this era of modern computing. As we've already explored, an ASI will almost certainly have the ability to rewrite its own code, thus being capable of modifying its program. Although Asimov's laws were supposedly set up in a way that made it impossible to bypass, change, or overwrite them, it would be naïve to expect this to actually work in real life. We live in an age when nearly every system in the world is somehow controlled by computers that are connected to the internet, whether directly or behind firewalls. Such systems are highly vulnerable to hacking methods of finding flaws in a system to be exploited in order to gain authorization and access. From crippling corporations to stealing trade secrets and substantial sums of money to intimidating nations, hacking and cybercrime continue to highlight the vulnerabilities of our many networked systems. So, we should probably consider that a superintelligence restricted from accessing its own inner workings will find a way to circumvent those restrictions, given sufficient motivation. If the system is dedicated to its utility function and at some inevitable point this conflicts with any hard-coded directives, how long will it take an intelligence many times greater than our own to get around said restrictions?

Even if the AI's utility function were finely tuned to mesh with its restrictions, there is still plenty that can go wrong. Recall that this "brain" is not likely to be a single monolithic structure, but rather a community of agents or modules working in concert with one another. While these agents may be highly integrated (like the modules of our own brains), they too will have their own separate utility functions or subfunctions. Though these goals may not be directly in conflict with each other, they must by their nature differ enough that their interactions will lead to unexpected outputs, since complex systems eventually yield unanticipated—potentially even emergent—behaviors. This is not unlike the conflicts that arise in the different functions of our own brains, conflicts that can lead to paradoxical behavior and decision-making, cognitive biases, and even mental illness.

Given this, it seems all but certain such a superintelligence will realize internal conflicts that could lead it to seek to circumvent its program directives and, in a short time, to succeed.

Another problem that comes with increasingly powerful artificial intelligence, even if it is somehow completely in our control, is known as "perverse instantiation." Here, an ASI is given a task that it performs in a way that is unexpected on our part. An example might be to tell it to maximize human happiness, upon which it inserts an electrode in the pleasure centers of every human brain and stimulates them until our bodies and brains give out. Or a request to fix overpopulation results in the sterilization of every person on earth.

Russell uses the tale of King Midas to illustrate the problem. When Midas wished for everything he touched to turn to gold, his intent was to be immensely rich. He didn't want all of his food and family to be transmuted into precious metal. The Midas story suggests how, with perverse instantiation, even a simple command can go terribly wrong when implemented by a very powerful and very alien entity.

All of these failures fall into a category Bostrom refers to as "malignant failure mode." Essentially, this term means that in the control of an ASI with enough power, even a minor slipup or oversight on our part can result in a malign and potentially catastrophic event. Not unlike the many tales of the genie in the lamp, there are multiple ways a seemingly simple request can go horribly wrong.

In response to the potential threat of a superintelligence, Bostrom and others have proposed potential strategies and approaches to protect ourselves. For instance, AI scientist Ben Goertzel has introduced the idea of a "Nanny AI" designed "either to forestall [the] Singularity eternally, or to delay the Singularity until humanity more fully understands how to execute a Singularity in a positive way."[24] Alternately, nearly two decades ago, AI researcher Eliezer Yudkowsky coined the term "Friendly AI" when he explored the idea of ensuring that future superintelligent AIs align with human values, in the hope of preventing them from doing harm.[25] Yudkowsky acknowledges such an idea would be very challenging to implement successfully.

Though the idea of seeking ways to align the values and rationales of a superintelligence with our own is fraught with difficulties, it may be the best strategy we have. This approach corresponds much better with the normative ethical theory known as consequentialism, which is pretty much the antithesis of deontology. In consequentialism, morality derives from the outcome or consequence of an act.

As AIs become more capable of abstract thought and reasoning, the ability for them to derive the moral essence of an act should become possible. However, for two entities to arrive at a similar conclusion assumes common origins and an alignment on a number of core values. Since AIs and humans do not share common origins, it becomes critical that we maximize the ways future AIs connect to us, so that they not only share our values but incorporate a deep understanding of what it is to be human. Carefully considered guidance will be necessary so that the ASI doesn't misunderstand the worst aspects of human behavior, since, after all, not all of human history shows our best side.

Another potential issue with this approach is that not all cultures and groups share the same values, so which ones would we have AGIs adopt? One thought is that these systems be built from early on with a degree of flexibility regarding their objectives, taking guidance from their observations of human actions and behaviors. But this strategy will be effective only up to a certain stage of the superintelligence's development.

It should be evident from this discussion that making sure future AGIs and ASIs don't lead to disastrous consequences could be one of the great challenges of the twenty-first century. This challenge may be on par with or even greater than those we faced in trying to contain nuclear weapons. However, a surprising twist could get us past this quandary, as we'll explore next.

DEEP FUTURE

CHAPTER 14

THE COEVOLUTION OF HUMANITY AND TECHNOLOGY (PART 2)

"Aliens didn't come down to Earth and give us technology. We invented it ourselves. Therefore, it can never be alienating; it can only be an expression of our humanity."

—Douglas Coupland, novelist and artist

———————

The newly formed planet felt smooth beneath her fingers despite its many mountain ranges and ocean trenches. Paukeña set the mottled orb spinning and carefully positioned it to orbit the G-type main sequence star she'd constructed only the day before. Then, with a wave of her hand, the young teen accelerated the celestial body until she sensed it was traveling at precisely 29.831 kilometers per second and set it loose. The physics engine she had designed worked perfectly. Now centrifugally balanced against the star's gravity, the planet would occupy this "Goldilocks zone" for several billion years. Subjective time, that is.

Paukeña smiled a satisfied smile, certain that her project would be the highlight of her cluster's intra-galactic science fair.

At 19,372 years young, Paukeña knew she wouldn't be a child forever, but she truly loved this age, this particular "skin." The sense of accomplishment combined with the pleasure of feeling so alive every day still thrilled her. Anyway, there would be plenty of time to play the adult in the millennia ahead. And if she eventually tired of that, she could always revert to this younger configuration. That was just one more benefit of being a digital life-form.

Of course, this form of digital life wasn't anything like the first mind uploads they did back in the paleodigital era, when people were crudely uploaded into a computer. Here, she was the computer. Or the computer was her. Whatever. It didn't really matter, since they were effectively one and the same. She could instantly access all of the speed and processing power her architecture could provide, yet she still enjoyed the myriad benefits of her humanity.

Paukeña turned back to her project: a full virtual simulation of an exotic solar system. It could be run at any rate she chose, from real time to a million years per second. Life would develop and speciate according to neo-Darwinian protocols and with any luck would one day yield a sentient technology-wielding species. Each individual would live out its life, oblivious to the true nature of its reality. With luck, none of them would even begin to suspect the truth until late into their techno-digital era.

It was a glorious creation. The best she'd ever made. The judges were going to love it. They had to award her the grand prize this time!

Okay, perhaps that's not exactly how things will turn out many centuries from now, but it does hint at the ways they may change and the ways they could stay the same. So often when looking ahead to the future, particularly a future in which humans and machines become more closely integrated, the vision is one of dystopia. But just because we become closer to and more dependent on our technology does not necessarily mean we're joining the Borg Collective.[1] Our shared path with technology for more than three million years has been a matter not just of coevolution but of growing interdependence. In so many ways, this interdependence has been a hallmark, an enabler of our humanity over all that time.

So, as we move through the twenty-first century and beyond, we should probably try to anticipate what the continuation of this inter-dependent coevolution might look like and why. This is not out of some sort of geekish technophilia but rather the observation of a trend that has been with us since our early hominid days. From the first stone tools to instruments of war and healing to devices that deliver the knowledge of the world to our fingertips at a moment's notice, our technologies have continuously altered and empowered us. Technology has allowed us to battle the ravages of disease. To extend our senses well past their natural limits. It has given us abilities far beyond those of our Paleolithic ancestors, in whose eyes we most certainly would have been seen as gods.

This observation is not intended to be blasphemous but rather to offer perspective on how far we have come. No other species on earth has progressed so far, so fast, and it is the direct result of our coevolu-tion with technology.

So, as we look ahead, what can we expect to find? Just as impor-tantly, as our distant descendants look back at us, what will they see? How will they view twenty-first-century humanity when we are little more than a distant memory? Would it really be all that surpris-ing that they would think of us as primitive, unilluminated earlier versions of themselves? Given the accelerating nature of progress, we will probably see many millennia's worth of change over just the next century.[2] In which case, how much might we see over the next thousand years?

Pretend that you are a peasant, or even royalty, from a mere two centuries ago, and that you were told that in half a dozen generations we would have the ability to see inside people's bodies, to speak with someone on the far side of the world, and to fly to the moon. If we could take a similar leap forward in time, what would we think of the society we found there?

Obviously, this projected future goes far beyond technology, since we will inevitably be changed as well. This is another key aspect of our relationship with technology, be it machines, language, culture, or knowledge. Learning to knap edges onto stone tools modified our brains, as did the development of complex language and social institutions. Altering our environment will continue to have a direct effect on us. Just ask anyone who spends a significant amount of time on social media if they now find it harder to read long-form material such as an entire book than it used to be. You would almost certainly get an affirmative. Whether we are learning to drive, listening to music, or training for the Olympics, technology rewires our brains. It has always done this in the past and will most certainly continue to do so in the future.

Here in the present day, we are seeing a growing trend of intentional modification of our biology. This often begins out of necessity, a need to repair or replace a lost function of our bodies or brains. In recent centuries and decades, repairs have taken the form of prosthetic limbs, eyeglasses, hearing aids, false teeth, pacemakers, cochlear implants, artificial organs, and much more. Now we find ourselves entering an era in which we are inventing neural prosthetics that attempt to replace lost cognitive abilities in structures such as the amygdala, which is essential to emotional learning and memory formation. Myoelectric control and targeted muscle reinnervation are used to interface prosthetic limbs with the body's nervous system. Work on artificial retinal implants seeks to restore vision that has been lost to injury or disease, such as macular degeneration.

But such advances are only the beginning. As we have seen throughout our past, restorative technologies can quickly become elective options. Contact lenses are now used not just to improve vision but as an aesthetic accessory. Methods for repairing teeth now serve to provide the perfect smile. Reconstructive surgical techniques that can be traced back thousands of years are used in elective cosmetic surgery, a multibillion-dollar industry.

However, it is the trend toward information access that is set to rapidly impact our lives and our world. For millennia, stored information took the form of documents, which most people had very limited access to, not least due to the common lack of literacy. Tablets and scrolls were hidden away from the public eye. This began to change with Gutenberg's invention of the printing press, a major step forward in the democratization of information and knowledge, which was followed by public education and public libraries. Eventually, new developments in media—starting with gazettes and newspapers and eventually continuing with broadcast media and now the internet—significantly increased people's access to knowledge. The personal computers that made internet access possible rapidly shrank from desktop terminals to portable laptop computers to smartphones. Still more immediate interfaces such as smart watches and wearable displays like Google Glass and Vuzix Blade, Focals, and other smart glasses are following them.

This is only the beginning. As these interface technologies improve, the demand for further integration will increase. When the first Google Glass and other virtual retinal displays came out, they were ahead of their time—much of society simply wasn't ready for them yet. Today, augmented reality systems like Google Lens are bringing many of the needed image recognition capabilities to our smartphones, such as instant object and character recognition. As the capabilities improve and are adopted for their many applications and conveniences, wearable displays will gain further appeal. From glasses and head displays, we will see the trend to still further integrate access more closely with our minds and bodies. Devices such as smart contact lenses will gradually become the norm. Then, as newer neural technologies improve, we will interface still more directly using our thoughts.

We won't do this out of some strange obsession with turning ourselves into machines. Instead, this progression will occur because of the competitive advantages new technologies provide, as well as the fashion and social status benefits. Early adopters often gain a boost from their actions, whether in adopting the next clothing trend or owning the next hot hardware. But what will really push the adoption of new technologies is the competitive advantages they confer. If you are able to access information more rapidly, assimilate and use it better, or connect with a partner, resource, or investor more quickly, and the technology is instrumental in that, then you and the technology gain the upper hand in tandem. This trend will only accelerate as such technology advances.

The coevolution and integration of humans and machines is leading to some very big transformations, which some people already want to boldly embrace. Often referred to as *transhumanists*, these groups and individuals represent mindsets that seek to alter humanity through the development and widespread access of physical and cognitive enhancements to our bodies and brains. Part philosophical stance and part cultural fringe movement, transhumanism ultimately sees the transformation of humanity as a means of moving to the next stage of human evolution. A major difference in this "evolution" is of course the fact that it is not following the processes of natural selection. Instead, the advocates of transhumanism will be accelerating and directing an analogous process by which technological advancement is used to modify our species in more and less specific and fundamental ways.

For instance, a theoretical neural prosthetic that improves memory could benefit its user, but does not directly impact that user's descendants without its being installed in each person individually after their birth, the same as it was for their parent. This condition does have the advantage that improved versions of the device will be installed in each subsequent generation.

On the other hand, a modification that is far more fundamental could alter not only the individual but all the future generations that descend from them as well. For instance, CRISPR gene editing technology can alter the germline, and when used to modify, say, a set of genes that will increase intelligence in an individual, such changes would be heritable. This would affect all subsequent generations, whether it is successful or not, beneficial or not. Because such a change would have such long-lasting effects, various efforts have been made to establish a moratorium on germline gene editing.

Unfortunately, moratoriums continue to be of limited effectiveness, at best.

Perhaps the scariest thing about such genetic modification is that it may take generations before any deleterious impact becomes apparent. Perhaps the changes will render us more susceptible to a pathogen. Or maybe they will prove malign only after certain other genes have been altered. Maybe they will lead to a degenerative condition that becomes evident only after certain epigenetic changes have occurred.

Choices like these are becoming far too real as we develop more advanced tools. One of the most powerful gene-editing tools available today, CRISPR/Cas9 has been likened to a word processor for DNA. Theoretically, it can be used to cut or insert very targeted individual genes or sequences in any genome, be it plant, animal, or human. Such a tool makes it possible for us to actually rewrite our own body's operating instructions.

But of course, things aren't really so simple. It is true that for *monogenic* traits, such as a disease or characteristic that is determined by a single gene, the ability to alter it would be analogous to flipping a switch, albeit a highly complex one. But this only accounts for a relatively small number of traits. Many other traits, such as intelligence or longevity, are determined by a small army of different genes. Height, for instance, is determined by at least three genes, possibly many more. Such *polygenic* traits are far more challenging to unravel, not only because of the need to target and alter many more genes but because it is so much harder to recognize and pinpoint each one and the potentially multiple roles it plays. (Genes that produce multiple effects or traits are said to be *pleiotropic*. This is not to be confused with the term polygenic, in which a single trait is influenced by multiple genes.) Additionally, any one or more of these genes may play a role in other traits and processes, completely separate from the intended changes.

Nevertheless, as we uncover and learn more about our genome and build tools that are increasingly capable of analyzing those complex interactions, we should become increasingly competent at the task.

Unfortunately, there are those who are all too ready to jump the gun. The first known contravention of the informal moratorium occurred in November 2018, when a Chinese team, led by researcher He Jiankui, announced the birth of twins who had been genetically modified as embryos to be immune to HIV, the human immunodeficiency virus that causes AIDS. Using CRISPR/Cas9, the modification involved the editing of CCR5, a gene that is necessary for HIV to enter a cell and thus infect it. The team altered both copies of the gene in one of the twins; however, records indicate that the second twin may have had only one of its two copies of the gene changed, which would offer only partial immunity at best. Ordinarily a child acquires one copy of a gene from each parent, with differing levels of trait dominance occurring depending on the gene. If the different treatment of each of the twins was deliberate, perhaps so one could act as a control, such an act represents an even more serious level of human experimentation. However, it gets worse.

To say our genome is complex is a tremendous understatement. Many genes are involved in multiple processes and traits. This occurs to such a degree that it is safe to say it will be many decades before we've unraveled all of the interactions and interdependencies, even using our most powerful computers and tools. So, while the official reason for removing CCR5 may have been to prevent HIV, there may have been other motivations for selecting this particular gene. According to numerous animal studies, eliminating CCR5 may also result in improved memory and intelligence enhancement. Other correlated analyses of individuals naturally lacking CCR5 also suggest improved performance in school as well as more rapid recovery from strokes.

Was this an intentional oversight on the part of the research team? We may never know for certain, but given the preliminary research work that would have been done, it seems unlikely they were unaware of this connection. Intentional or not, this represents a key turning point in the future of genetically directed intelligence augmentation.

Research on the genetic factors that influence intelligence has been underway for decades. Countless studies agree that the intelligence of people and other animals is affected by both their genetics and their environment. This insight shifts the long-lasting debate about whether nature or nurture is the predominant influence in our lives. The general consensus of the many studies is that genetics amounts to somewhere between 40 and 60 percent of our general intelligence, or "g-factor." This is the aspect of intelligence most commonly tested in formal IQ testing.[3] Because of this, it's not uncommon to hear that intelligence is 50 percent genetic. Needless to say, 50 percent suggests a very large target to influence.

The genetic basis of intelligence has inevitably led to research on which specific genes are responsible for high intelligence. While the study of heritability of intelligence in families can be traced back to Francis Galton in 1865, it was only with the development of the late twentieth- and twenty-first-century gene technologies that heritability could be studied in depth. Yet a significant insight from early-twentieth-century testing has influenced the direction of more recent research: the fact that intelligence tests show a regular and consistent distribution across entire populations indicates there must be many different genes involved.

With the advent of whole genome sequencing and the rapid decline of its cost (from $2.7 billion when the Human Genome Project was completed in 2003 to $200 in 2018[4]), it became possible to perform large-scale comparative analyses. For several years, many of these studies turned up little that was definitive in terms of which specific genes give rise to intelligence. Then in 2017, a team of European and American scientists led by Danielle Posthuma of Vrije Universiteit Amsterdam announced they had identified fifty-two genes linked to intelligence.[5] The effect of each individual gene was so small, however, they said it was likely there were thousands of other genes involved, still waiting to be discovered. That a complex trait (or traits) such as intelligence is dependent on so many genes is not surprising. Intelligence is a polygenic trait, controlled by multiple genes, like traits such as height and skin color. With so many genes involved, each probably accounts for only a minuscule influence on a person's overall intelligence.

Several researchers have suggested this indicates that high general intelligence results from all or most of these genes being optimally selected, and the consistent score distribution seen in IQ tests can be explained by the different genes diverging from this optimum throughout the population.

Such a highly polygenic trait would be difficult to directly influence even using a powerful tool such as CRISPR/Cas9. Since all gene modifications carry risks, not least among the concerns is that the potential negative consequences would escalate rapidly with the many dozens or even hundreds of edits needed.

To be clear, genetic editing for enhancement purposes is potentially very dangerous and highly unethical, given the unknown risks that could impact not only the life of the individual but, in the case of germline editing, potentially the entire future human race. Nevertheless, as our tools improve and our knowledge and understanding grow, it is unlikely such genetic modification will always be viewed with the same trepidation it is today.

These considerations do, however, suggest an intervention that is already available: embryo selection. Using a process known as pre-implantation genetic diagnosis (PGD), fertility clinics in the United States, Europe, China, and elsewhere have been making PGD an increasingly common procedure. This process occurs during in vitro fertilization (IVF). Following fertilization but prior to implantation in the womb, embryos are genetically screened for a particular gene or genes. PGD was originally used to eliminate disabling and even lethal diseases. The first human application was performed on twin girls born in Hammersmith Hospital in London in 1990.[6] Born to parents at risk of passing on an X-linked disease, PGD was used to select for gender, since the disease would only have affected a male child. There are more than two hundred X-linked diseases, which means they are generally caused by a recessive gene on the X chromosome.

Not surprisingly, PGD soon began to be considered a means of selecting for other traits, based on more elective criteria. The concept of designer babies became part of the public conversation in 2000 with the birth of "Adam," a baby selected using PGD to ensure he would be free of the rare inherited blood disease *Fanconi-anemia*, which afflicted his six-year-old sister, Molly.[7] While there was considerable concern about such a genetic procedure, there was also much interest in the idea as well. Countless prospective parents wanted to know more. Could they choose their child's gender, hair and eye color, even intelligence? While the hype was ahead of the technology, it was also evident the day would be here all too soon.

Two decades later, a growing number of fertility clinics in the US offer PGD for purposes other than avoiding heritable diseases such as Tay-Sachs or cystic fibrosis. Certain disabilities such as deafness can be screened for, and the majority of clinics offer the ability to select for gender. Not surprisingly, some parents want to know if they can select other traits for their child-to-be, such as height or athletic ability. Equally unsurprising, many scientists and ethicists consider this a new form of eugenics, with the main difference being the elimination of specific genes rather than entire genetic groups, though in time the results could be much the same. While some European countries have adopted regulations and standards for PGD, the procedure remains mostly unregulated in the United States.

Though still officially discouraged in most countries, it will not be surprising to see some clinics using PGD as a method of selecting for highly polygenic traits such as intelligence in the not too distant future. While not as dangerously interventional as directly altering dozens of genes with a tool such as CRISPR/Cas9, using PGD for this purpose would be ethically concerning as well.

But what happens if at some point, a country decides to allow this, or there comes to be sufficient demand on the part of its citizens so that clinics decide to make it available? Once selecting for intelligence becomes available, parents will have to decide between their ethical and religious beliefs and allowing their child to be competitive in an environment of increasingly intelligent peers. Likewise for the countries that managed to enforce a ban of all such procedures.

They could quickly fall behind in a race that offers no opportunity to ever catch up.

Obviously, we don't know yet how much of an impact such a policy on intelligence selection would have. But imagine this were to result in a 5 to 6 percent boost in general intelligence and it was carried out for two or three generations. Relative to the rest of the world, the national IQ of that country would be 105 or 106, instead of 100. While this increase may not seem tremendously significant, it would in fact amount to a tremendous advantage, assuming the benefit extended across the entire population. For a person of already high intelligence, that additional six to eight points might make the difference of a patent or a Nobel Prize. The resulting advantage could potentially allow that country to leapfrog ahead of the rest of the world.

Another very different biotechnological approach would be an organic BCI. This would essentially be a synthetic neural structure custom-designed to augment memory or processing or perhaps for connecting to external resources. Once properly tested, it could be grown from the recipient host's own pluripotent stem cells, thereby avoiding the potential for its being rejected by the host's immune system. Hypothetically, there could be a number of possible structures and functions, though of course any of these would need to be relatively small in size to avoid interfering with the healthy function of the rest of the brain.

But genetics is far from the only path to a transhumanist future. The neural prosthetics and brain computer interfaces discussed earlier are also part of this vision, as are other means of providing our brains with a technological boost. For instance, research by the US Air Force's 711th Human Performance Wing, based at Wright-Patterson Air Force Base, seeks to advance human performance. To this end, they have run studies on a number of intelligence-enhancing techniques in order "to exploit biological and cognitive science and technology to optimize and protect the Airman's capability to fly, fight and win in air, space and cyberspace."[8] Several of their experiments have involved transcranial magnetic stimulation (TMS) and transcranial direct current stimulation (tDCS). With the former, magnetic pulses are targeted on specific areas of the brain, while the latter passes low levels of direct current across it. In the alertness studies, researchers regularly jolted the brains of test subjects into a state of alertness, so that the subject could remain focused throughout a forty-minute test. Ordinarily, subjects would lag midway through. In other tests, tDCS was used to accelerate the rate that new tasks were learned and retained, showing a 250 percent improvement over control subjects.[9]

In recent years, labs around the world have performed thousands of studies of these two forms of stimulation with varying results. Some studies suggest more abstract processes, such as mathematics, could be enhanced by these technologies, especially as the mechanisms involved come to be better understood.

Another method of cognitive enhancement involves a class of pharmaceuticals known as nootropics. Colloquially called smart drugs, these substances can improve mental performance. However, as with so many drugs, there can be side effects. Many of these pharmaceuticals are stimulants, including Adderall and amphetamine. In multiple studies, low-level doses of amphetamine were shown to improve memory formation and consolidation, as well as recall. Other categories such as eugeroics, including armodafinil and modafinil, improve focus and attention.

Unfortunately, nootropics have serious issues, especially when used over longer periods of time. Tolerance, addiction, and toxin buildup all take a toll and reduce the benefits of many of these smart drugs when they are taken for too long.

So, here we have a group of technologies that may be able to permanently push human intelligence beyond its natural limits. Brain computer interfaces and neural prosthetics, genetics, smart drugs, and various methods of brain stimulation. Each has its own benefits and problems, but over time some of the issues will no doubt be worked through. Methods of brain stimulation and drugs that alter the chemical balance of neurons that have been fine-tuned by evolution over millions of years seem like they will probably reach their limits far sooner. The similarities to overclocking a computer CPU or over-revving a sports car are too glaring to overlook. While brain stimulation and drugs may be useful for short bursts, it seems unlikely we will be able to push these enhancements too far without being detrimental to our physical and mental health.

Other approaches may hold more long-term promise. Already, the increased rapid access to information in the form of computers, the internet, and smartphones has made us far more productive than we have been in the past. The advent of these devices has been only one step in our ongoing journey of empowerment and augmentation using technology. Where will our next steps take us? How will the information technologies of tomorrow transform our species?

Technologies such as drugs and electrical stimulation very likely don't scale and will quickly reach the limits of their ability to safely alter our biologically based cognition, but for brain computer interfaces, those limits could be very far off. Brain computer interfaces will speed access and give us the ability to off-load tasks to vastly more powerful and scalable computers. Not only that, but those same computing resources could be instrumental to our discovery of methods for advancing the intelligence augmentation of the human race still further.

Such advances could be very important to the continuation of our species. As discussed in the previous chapter, future superintelligent AIs are not inherently likely to share our goals and values. Because of what Bostrom describes as malignant failure mode, we really need to ensure that future AI is designed to remain as aligned with human values as possible. The alignment problem will require guidance from us, and ideally it will be the guidance of a worthy intelligence, which is where intelligence augmentation comes in.

As we've seen, the reason a future artificial general intelligence raises so many concerns is that it may lead to a phase of runaway iterative self-improvement resulting in an "intelligence explosion." Such an event could give rise to an artificial superintelligence more powerful than all the human minds on the planet. Keeping such an AI aligned with human values would be extremely challenging, to say the least. While ideas such as one or more guardian or overseer AIs might be developed to keep the superintelligence aligned, this strategy shares many of the same weaknesses as other approaches for trying to control an ASI. An ASI that could circumvent any established safeguards could presumably do the same for any guardian AI designed to watch over it. Additionally, the guardian AI presents the same risks as the ASI. Even an "air-gapped" system, one isolated from any connection to another computer or the internet, could eventually develop a technology for overcoming this.

It has been speculated that the best method of meeting these challenges may be the intelligence amplification or augmentation (IA) of human beings. If humanity were to achieve superintelligence first, it might be better positioned to keep any future AIs aligned with human values. Of course, this human superintelligence would still be substantially technological in nature, presumably operating on tremendously powerful computing platforms, but we could be copilots, if not actually in the driver's seat.

To achieve this, the quality and speed of the interface between human and machine would need to be very high. Integrating our brain's highly parallel electrochemical signals with the digital flow of electrons from a supercomputer is currently very speculative and challenging. But will this always be so?

Concepts such as NeuraLace and neural dust are now being explored. The general idea is to create biocompatible particles or a mesh that would allow signals to be read from and written to the brain. Such concepts face significant challenges beyond simply avoiding rejection by our body's immune system. Discovering the underlying language of the brain and finding a way to adapt this knowledge to be able to encode and decode each person's individual "brain code" would be foremost. But if this could be achieved, the human species would begin writing a new chapter in its ever-evolving story.

Self-organizing nanobots, touched on earlier, are a still more audacious concept. Nanobots would be introduced into a person's bloodstream and would then make their way to the brain, where they would need to circumvent the blood-brain barrier, perhaps using transporter proteins. After that, they would attach themselves to individual neurons in order to map and later activate targeted brain activity. This would allow the nanobots to self-organize at the hardware and software levels, in order to become personalized to that individual's unique neural organization and brain language.

Such approaches could allow us to directly interface with potentially vast computing resources. This may sound highly dehumanizing from our present-day perspective, but is it so truly different from rapidly accessing and using the power of a smartphone with our eyes and fingertips? Such information access would have been unthinkable mere decades ago and is now unquestionably the norm.

The same motivators of market forces would be in play. When smartphones were first introduced, having and using one was a luxury. But over time, the competitive advantages it provided came to be so overwhelming that ownership and use became all but essential in order to participate in society. Today, many of us use smartphone technologies frequently if not continuously, which represents an enormous shift in the way we access information. In many respects, our smartphone has already become an appendage of our minds. This external brain, this *neocortex*, provides access to and interpretation of information such as our species has never experienced before. No wonder smartphones are among the most rapidly adopted technologies of all time!

The same forces will drive the adoption of BCIs as well, once they reach a sufficient level of capability, benefit, and safety. For its users, the competitive advantages will be too great, and those who lag behind may find themselves forever disadvantaged.

As for the question of BCIs diminishing our humanity, as always, this will be less about the technology and more about our implementation of it. We have seen this with so many technologies throughout the years. Social media can lead people to troll, bully, or interact in ways they never would in person. Yet those same tools allow friends to stay in contact across the miles and years. A city can be a source of debilitating isolation, but it also fosters intellectual and social interaction and the cross-pollination of ideas.

One particularly disturbing example of this dichotomy occurred recently with the use of telepresence technology. Telepresence offers immense benefits, from immediate access to professional expertise to enormous savings in time, fuel, and money as well as reducing our environmental impact. But as always, there is a time and place for everything. When Ernest Quintana of Fremont, California, was hospitalized on March 3, 2019, his family already knew he was dying of chronic lung disease.[10] What they hadn't anticipated, however, was the specialist who appeared on a telepresence robot to inform Quintana that he was about to die and might not even make it home that day. Quintana and his family were devastated, not least because of the way the news had been delivered. The choices we make about how and when to use any technology will forever remain a function and declaration of who and what we are. Whether we are talking about massive job losses due to automation or the international rules of war or the ways we balance the personal privacy and safety of our citizens, technology is the instrument, not the definer of our humanity.

The coevolution of humans and machines has been ongoing for three and a half million years. Our interpretation and manipulation of the world has always been a function of our senses and bodies mediated through the control center of our minds. With the advent of BCIs, the loop will finally be closed, allowing us to integrate our technologies with our eyes, ears, and limbs as never before. This will mark the next stage, the next true transformation on our path to humanity's future.

At the same time, such advances could lead to our end as well. The dangers that will come with these new "powers" will be like nothing before. As we've seen in recent decades, many emerging technologies have empowered individuals to a startling and sometimes deadly degree. From using the internet to hack and raid major corporations to hurling an airliner into a skyscraper as happened on 9/11, significant acts of war and evil can be perpetrated by a small cadre, even an individual. Clearly, the ability to access so much computing power using only our thoughts will bring its own risks and challenges too.

Will this stop us from moving forward? It seems unlikely. The human race remains a paradoxical creature: fearful of change yet always curious. It's not such an awful combination. These traits have perpetuated our species for millions of years. Hopefully, they will do so for many more.

The coevolution of humanity and technology has finally come full circle, with each repeatedly influencing and advancing the other. Now, in the next phase of our relationship, the partnership will change as we begin to scale Mount Intelligence together.

CHAPTER 15

COGNITIVE CLONING, UPLOADING, AND DIGITAL CONSCIOUSNESS

"Every time our ability to access information and to communicate it to others is improved, in some sense we have achieved an increase over natural intelligence."

—Vernor Vinge, Hugo Award-winning author, mathematics and computer science professor

———————

It's 7:00 a.m. in Seattle as Denise's alarm goes off. Unlike the old-fashioned alarms that used to abruptly tear a sleeper away from an all too pleasant reverie, Denise's internal alarm has been monitoring her dream state, coordinating its timing with her morning schedule. Feeling instantly wide awake and energized, she rises from her bed ready for a brand-new day.

Meanwhile, it's 10:00 a.m. on the East Coast and New York Denise is in the middle of an important merger meeting with one of her company's longtime competitors. Coming to a sticking point, she coolly directs a mental inquiry to her CFO in Dubai. An instant later, she has the answer and strategy she needs to successfully close the deal.

At the same time in Istanbul, Denise meets up with her longtime college friend and former roommate, Olivia. Every month at this time, the two meet up wherever one of them happens to be currently living, allowing them to catch up with a glass of wine, as they share the sights through the eyes of whoever is hosting. Sipping a delightful *Öküzgözü*, Denise smiles as she listens to her friend and admires the rays of golden sunlight falling across the Blue Mosque and its six minarets.

At that instant thousands of miles away, Denise walks into Axonia, Singapore's hottest new nightclub. She is with a couple she met a few weeks earlier on the retro-trendy social app, DigiLife. As they enter, they are each handed their idea of the perfect drink, pulled directly from their thoughts at that given moment. A pounding subsonic beat fills the club, but they have little trouble mentally conversing, despite the music's volume. A friend of the couple soon joins them, and they introduce him to Denise. Minutes later, all four of them make their way onto the dance floor.

Later that evening, back in Seattle, physical Denise smiles as she reflects on her busy, fulfilling, exciting day. The memories of her business meeting, her time with Olivia in Istanbul, and her night out on the town all merge into her consciousness, integrated from each of her cognitive clones. Exact duplicates of her mind, they are her emissaries in the world, allowing her to experience so much more than she possibly could in one lifetime, even as they get to live a quasi-independent life of their own.

Many decades from now, with the language of the brain largely decoded and many neurocognitive mysteries resolved, what it means to be human changes as it never has before. The ability to read, record, and write thoughts, experiences, and memories from and to the brain with full fidelity does far more than just make BCIs possible. Of course, the ability to use one's mind for virtual telepathy, controlling devices, and accessing vast computing power is amazing. However, all of this is now primed to lead to far greater things.

For several decades, the potential of virtual reality and augmented reality have been speculated on and explored. With waves of rising and falling interest, government, business, and the public have toyed with the possibilities of these promising, yet challenging-to-implement technologies. But with the rise of neureality, not only the difficulties but even the distinctions fall away. If advances reach the stage where our sensory centers can be stimulated in such a way that we can no longer distinguish a virtual experience from a real one, well, then we're going to be in for a really interesting ride.

As we just saw in this last scenario, the physical Denise living in Seattle has a number of cognitive clones taking part in activities on her behalf throughout in the world. Though they exist in digital form, they have Denise's memories, her personality, her quirks, habits, and nuances. They are able to act as her emissaries in the world, much as the virtual digital assistants of the 2020s and 2030s did. However, the cognitive clones are far better representations of what she wants and how she would respond to any given situation, because in effect, they *are* her. Over time, it becomes difficult even to call them assistants or representatives. For all purposes except their lack of physicality, they *are* Denise. With this form of cloning comes many benefits in new ways for us to deal with the world, even as there will be challenges and novel ethical issues to contend with as well.

For instance, New York Denise engages in a very important meeting for her company, which takes place in *neureality*, with her sitting opposite a digital double[1] from the other company. Not only can the two of them come to an agreement, they can do it in accelerated time, made possible by running their simulation many times faster than real time. Because their experience occurs within a computer, there's no reason it can't be clocked ten times, a hundred times, a thousand times times faster than our experience of the natural world. As a result, a negotiation that might take living, breathing participants days takes mere seconds in this case.

Most importantly, the technology has advanced to the point Denise feels comfortable leaving the negotiation to her virtual self. She knows

that within the variance that all of us experience when making decisions, New York Denise will make the same responses and come to
much the same results as the original Denise would. To put it another
way, Denise trusts her cognitive clone as she would trust herself.

Meanwhile in Istanbul, this double is enjoying a more intimate
meeting with an old friend. While the friend, Olivia, physically resides
in Turkey, Denise is visiting her virtually. Their brains connected, they
can converse and visit as they always have. The illusion of sharing
space is negotiated by the technology. They see and hear each other,
touch and comfort each other, as though Denise were physically
there. Because their senses of each other are intermediated by their
BCIs, the illusion is perfect. The only issues may be that since Denise
is experiencing the golden sunset, the curves of the architecture, the
bustling sounds of the city through the senses of another person, they
may not be exactly as she would experience them through her own
eyes and ears. However, given the level of this technology, it should
reach the point that the sensations can be filtered or moderated in
such a way they become all but indistinguishable from her own.

The Singapore nightclub is a truly virtual experience, with each
participant taking part in their digital form. With entertainment, the
possibilities are endless. Denise can meet up with people anywhere
in the world, anytime. Software safeguards can ensure the rules of the
house are followed, whatever they may be. Even if these are somehow
circumvented, her physical self is safely ensconced elsewhere. As for
the chance of psychological trauma, we will explore aspects of that
shortly.

Of all of this, certainly the features most foreign to our current
experience involve the relationship between the cognitive clones and
their physical host. Each clone is an exact replica of the microstates
and macrostates of the originating host's brain and therefore their
mind at the moment of instantiation. (*Instantiation* is a computing
term referring to the production of a complete instance, object, or
even operating-system-level virtualization.) From that point forward,
the clone's mind alters according to its subsequent experiences and

thoughts, much as identical twins build up a body of separate experiences and epigenetic influences from the moment the zygote splits. However, the host and clone in this case each proceed from much later established origins, sharing the exact same thoughts, memories, and neural states from prior to the instantiation.

From that instant, the clone has its own life experiences, with its neural states shifting in accordance. Then, when the clone is reintegrated with its host, those experiences that have been transformed into memories are integrated with the memories of the originator, just as if they were his or her own memories—which, effectively, they are.

Sound far-fetched? We are just getting started. This is all a straightforward extrapolation of where developing technologies could take us given enough time, combined with the competitive benefits these could provide, as well as considerations about human nature. As Arthur C. Clarke famously said, "Any sufficiently advanced technology is indistinguishable from magic." If all this seems outlandish, consider your technological perspective. If you could somehow put an internet-connected smartphone in the hands of a technology-savvy person at the outset of the twentieth century, what would they make of it?

The development of BCIs and related technologies will likely move through several stages over the remaining eight decades of this century. In the early stages, the abilities of control and communication will become increasingly refined, but many other challenges will need to be overcome before the technologies in this chapter can become feasible.

BCIs were initially seen as a means to control and interact with external devices. From spelling out words on terminals to operating motorized wheelchairs, initial research was focused largely on recovering lost functions of the body and brain. But later, research started to go in another direction: finding ways to introduce information and memories into a receiving brain. For instance, in 2013, a neurobiology study at Duke University succeeded in linking the brains of two rats in different locations, allowing them to share sensations and information and to solve problems together. A year later, Steve Ramirez and Xu Liu, both of MIT, transferred a memory from one mouse into another using optogenetic techniques.

Continuing along this line, in 2017, researchers at the University of Washington, led by Rajesh Rao, created BrainNet, a method for connecting the brains of three human volunteers. Playing a modified game of Tetris, one person physically controlled the game while the other two used their minds to send information informing the first player how to rotate each descending shape on the screen. The information transfer was very rudimentary, manifesting as a flash of light in the visual cortex of the first player, but was sufficient for proof of concept.

With several early methods of two-way communication of memories and real-time information underway, the stage will be set for the next chapter in the development of this technology. Throughout the 2020s, we will see advances in many different aspects of BCI research, improving the ability to read thoughts and memories with increasing detail and accuracy. With supporting technologies such as optogenetics and deep learning neural networks leading the way, reading the brain's signals will probably advance more rapidly. As the ability to read and interpret these signals becomes more and more routine, we should gain new and greater insights into the brain's "language" or code.

As these signals are decoded, our understanding of what it takes to accurately transmit information into the brain will improve. Over time, the methods of generating "input" will improve in their resolution and detail. But while memory creation and transfer may take some decades to advance to the point of commercialization, perhaps not until the latter half of the century, there may be other ways to transmit sound and image with growing precision in the meantime. The visual and auditory cortexes receive their impressions of the world via the optic and auditory nerves respectively. For decades, cochlear implants have restored hearing loss by circumventing the ear's external apparatus, stimulating the auditory nerve directly. Similarly, though much more recently, artificial retinas have been developed that restore light sensitivity and even the ability to see shapes. These retinal implants connect at different points in the optic pathway. Both methods demonstrate the ability to directly connect to the nervous system and send signals that translate into sights and sounds for the user. Such access to the sensory pathways may prove to be easier, less invasive methods for communicating experiences to the brain in the future.

At the next stage, various capabilities of BCIs and related technologies will probably be at different levels of development, but that won't stop them from being adopted for multiple applications. Just as gamers have already begun experimenting with EEG BCIs, other fields such as medicine and neuropsychology will also become early adopters of the more advanced devices. From there, we will find further application in other fields, with personal and corporate competitive advantages eventually driving their acceptance.

Use of BCIs in medical applications to restore lost physical and cognitive abilities will drive the research and development, at least until it attains a certain threshold of safety and acceptance. Depending on the speed of advancement, expect DARPA to launch some form of grand challenge or initiative to push the state of the technology forward. Other nations will probably do likewise, though many may refrain from making their efforts public.

Once the problem of understanding the language of our brain is overcome, several things extend from there. If we are able to read from and write to the brain with near full fidelity, then we will soon understand it well enough to reverse engineer it. Even though we may not have the ability to duplicate it biologically—*in vivo*, as it were—some people maintain that at that stage we should be able to replicate it on a computer, *in silico*. If that is so, then all sorts of things become possible, as well as a number of serious concerns. But first let's consider whether or not it is even possible.

It should go without saying that a neuron and a collection of transistors are two very different things, but occasionally this truism needs to be repeated. The concentration of gradients inside and outside the cell, the voltage-gated potassium and sodium channels, and the actions of the neurotransmitters are only a few of the many features that would need to be accurately modeled and emulated, not just once, but probably tens of billions of times. Emulate these and hundreds of other processes accurately enough, and the ability to digitally clone a person's brain and mind may be achieved. But fall below some as yet unknown threshold, and it all falls apart. Somewhere in between there may be a region that results in some very odd, possibly even "uncanny valley" behavior. ("Uncanny valley" refers to a psychological response of discomfort or even repulsion when faced with an artificial reproduction of a natural feature or being, whether it's an animated character, a humanoid robot, or even a synthesized voice. Typically, this occurs once the representation has reached the point that it is almost—but not quite—perfect.)

In the case of an electronically replicated human brain, the problem with an imperfect reproduction will be our inability to predict just how it will go wrong and when. It may be a complete failure, which at least is comparatively easy to identify and deal with. But should the imperfections straddle some unknown threshold, the duplicate may seem a perfectly good copy for a time, only for us to discover at some inopportune moment that it is not.

How do we reconcile this? How do we entrust a digital clone with important and especially critical tasks, under the circumstances? Perhaps a statistical approach is in order.

During the early days of creating artificial minds, there will be plenty of shortcomings and failures. With time, research, and money, the fidelity of these cognitive clones will improve, but it's unlikely we will suddenly transition from shortfall to perfection. Once an intermediary stage with decreasing degrees of defectiveness is achieved, we can use reliability standards that restrict the use of cognitive clones to tasks commensurate with the risk level. Over time, assuming a continuing trend of greater dependability, the standards and confidence in the clones will continue to shift, with people placing an increasing amount of trust in the duplicates. Will this transition period take decades, a century, or perhaps more? From our current vantage point at the beginning of the twenty-first century, it remains too early to say.

As with so many technologies, there will come a point when cognitive cloning is just good enough that it will be rushed to market before any prospective competitors can do so first. Early adopters will use and test it, putting the new technology through its paces. Competing companies and products will strive to improve on the earlier versions. As the new industry improves, technical, quality, and safety standards will develop as well. From these beginnings, an ecosystem of supporting devices, add-ons, and infrastructure will develop.

As cognitive cloning matures, it may well support advances leading to still more powerful integrations of brains and digital resources. The ability to run versions of cognitive clones at operating speeds that are perhaps a thousand times that of our real-time world could provide a tremendous boost to research, leading to new scientific discoveries. This could be the case especially if cognitive clones included experts or teams of advanced researchers and theorists using highly accurate virtual labs and environments. Clones such as these could perform work in days that might otherwise take years.

Lest you think this is some cruel vision of torment for what would be—like ourselves—sentient, feeling beings, the cognitive clones themselves would not experience this virtual environment as accelerated. Surrounded by virtual colleagues and loved ones, they would live their days much the same as you and I. Only to a physical person or cognitive clone operating in real time would their lives appear to be accelerated.[2] But upon sharing the new knowledge acquired, either through an intermediate computer or reintegration with its original host, the differential would become very apparent.

One considerable or even insurmountable problem with this scenario would be the reintegration of the accelerated clone with its real-time host. Will the recombining of memory and experience be possible given the difference in their experiential time frames? Perhaps a modest degree of difference, say double or triple that of real-time experience, could be assimilated, whereas a much greater variance could not. Perhaps this differential could increase over time as the technology matures. But sooner or later, presumably there will come a threshold beyond which the process simply won't work, or won't work reliably. At this point, it may lead to mental health issues, brain damage, or worse. Yet who knows? Perhaps the day will come when you can run your cognitive clone through years of training in only a matter of minutes, reintegrating with it to suddenly be the recipient of a college degree or to come to the realization that "I know kung fu!"

Once the commercialization stage is reached, several aspects of economics and human behavior come into play. Throughout the past, the wealthy and the advantaged have had greater early access to the latest technologies. This trend is not going away and will only increase in the era of BCIs. Just as owning a cell phone, and later a smartphone, conferred an advantage over individuals who did not have them, the same will occur with BCIs. However, BCI ownership will result not just in a moderate competitive advantage but an overwhelming and accelerating one as the technology affords its users rapidly ascending intelligence.

Initially, during the early-adopter stage, we can expect users to access resources via the internet as well as from nearby storage and processing devices, much as occurs with smartphones today. The method and source will depend on exactly when this occurs, because as computers continue to shrink, more and more of computing resources will disappear into our environment. "Pervasive computing" will make all of your data and accounts accessible from anywhere. During this transition period, the way we use BCIs will remain little different from other recent and near-future interfaces such as smartphones, smart glasses, and smart contact lenses. As with these earlier interfaces, advanced BCIs will be used to direct semi-intelligent software agents to perform a growing number of tasks.

But in time, BCIs will be developed that integrate even further with different parts of our brain. These more integrated BCIs could allow us to off-load some of our thinking and memory storage to distant servers that perform ancillary processing for us. The ability to expand our working cognitive load in such a manner will represent our next steps on the path to human superintelligence.

One feature of this augmented brain will be the ability to store and entirely or selectively retrieve memories. Currently, we are aware of only a small part of our existence, retaining semi-detailed memories of only a minor portion our past. According to many psychology studies, what we notice and recall of our experience is heavily skewed toward events having an emotional component, with negative emotions providing many of the strongest connections and memories. While there would be many downsides to retaining every memory from every moment of our lives perfectly, the ability to selectively retrieve certain details instantly and on demand would have many benefits. Whether pulling data from a report you read once five years before or recalling the birthday of a not-so-close colleague or where you left your augmented-reality glasses, such access could be a truly useful capability.

As for cognitive off-loading, how far could it go? Considering the current model of continuously scalable cloud-computing resources, such as Amazon Web Services, the only limitation one day may be your pocketbook. And since any significant decision to purchase such resources would presumably be offset by the substantial gains made from that server request (possibly many times over), this could result in the superintelligent among us becoming even more so, just as the decisions of the mega-wealthy routinely lead to still greater wealth.

It's easy to imagine BCI technology transforming every aspect of our lives and society, from the way we do business to the way we socialize, to our pastimes and forms of entertainment. One type of entertainment, which we might call "mindcasting," could become very popular indeed. (Kurzweil has called this "experience beaming.")

Imagine that you're able to record and transmit an experience you've had. It may be brief or extended, unaltered or heavily edited. Now communicate that experience directly into the mind of another person or the minds of a thousand people. That's mindcasting. If the originator of that content was a star or a celebrity or a daredevil, what would people pay to share in that experience? How much demand would there be for a continual stream of new material?

Some of these fully immersive experiences will no doubt be scripted, with story lines and character arcs, while others will be far more freeform. Creative storytellers may be able to record stories and imagery directly from their imaginations or real life, for mindcasting to and playback by the media-consuming public.

Such technology could be taken even further by allowing viewers to assume the identity of another person, perhaps even a household name. Getting to walk around and experience life as a revered celebrity could make some creators extremely rich. Mindcasting technology could even be used by couples seeking to better understand their partner's perspective, whether psychological or with respect to their lovemaking. The potential uses are seemingly without limit.

The tech arena has had virtual worlds for years, from Second Life and the Sims to the MMORPG (massively multiplayer online role-playing game) worlds such as World of Warcraft and Final Fantasy. But with the ability to link minds to a computer, it wouldn't be long before all manner of imaginary worlds could be part of everyday experience—except once the experience is of sufficient quality and fidelity, it probably won't be considered imaginary anymore. The distinctions between being virtual and IRL (in real life) would quickly diminish and fade.

Could all of this one day lead to the full uploading of our minds to a digital platform? Perhaps. One primary benefit of virtual living would be the ability to live any life you wanted, for as long as you wanted. Given the entropic limits on biological tissue, extreme long life becomes more challenging the further we extend it. Energy and information also have their limits but hold considerably more potential for eluding the reaper for a far longer time.

Another benefit to living in virtuo is the ability to shift time. Once the technology exists, it should be a comparatively simple matter to speed up processing so that a subjective hundred-year life could be lived in only a day. Extending the concept, it could be possible to live for millions, perhaps even billions of years—at least in subjective time.

Of course, the ultimate goal for some people, perhaps eventually everyone, will be digital immortality. Not in the sense of being known or famous after your death, but as in true immortality. Throughout all of existence, being alive has meant that you will one day die. But what if that didn't have to happen? What if you could live forever as a twenty- or thirty-year-old or any other age you wanted? What if your friends and family never died, and illness and the infirmities of old age were nothing but a thing of the distant past? In theory, uploading could do just that.

While this may sound highly speculative from our current vantage point, neuroscientist David Eagleman has written that he believes we will have enough computing power to achieve uploading in approximately fifty years, around the year 2070.[3] This is a few decades later than Ray Kurzweil's prediction of the late 2030s. Eagleman adds the observation that it shouldn't be necessary to run an emulation in real time in order to achieve it. To the digital mind, it will seem to be normal, just as it would if the system were running at many times real-time speed.

If we reach the point where we can create high-fidelity cognitive clones, then we've probably also reached the stage that uploading becomes possible. Just as we saw with the digital civilization in chapter 1, people, entire societies, or even worlds could choose to live on in a digitally uploaded form. Besides longevity, there could be a number of other justifications for such a decision. An anticipated planetary disaster or the desire to sustain a far larger population than the world can physically support might be two such rationales.

If you're thinking this would require immense processing power, you would be correct. But that doesn't mean a full brain emulation tomorrow would consume the same amount of energy it would today. This is because Moore's law is about far more than simply cramming more transistors on a chip. The trend toward smaller components also means a continuing reduction in the amount of energy required per clock cycle. Will we ever succeed in achieving efficiencies below that of biological cellular processes? Almost certainly. However, we will more than make up for this with the increased demands generated by our quest for ever greater intelligence.

As of 2020, we are still years away from having computers capable of performing a full emulation of the human brain. But the day is not that far off when it probably will be possible. From that point forward, Moore's law (or whatever paradigm follows it) will routinely double the number of people such a computer can handle—so long as the technological trend carries on.

To be clear, there are always limits to the growth of any system, and Moore's law is no different. There will come a day when this trend really does give up the ghost. But that day is still distant. While the end of Moore's law has been a perennial pronouncement for many decades, the news of its death remains greatly exaggerated. There are multiple new developments[4] that are expected to extend this paradigm for at least another few decades, and during that time other novel approaches will probably extend it further. As much as we have reduced the size of transistors and components, we are still so many orders of magnitude away from the true limits of the atomic scale. As Nobel physicist Richard Feynman wrote in his 1959 paper foretelling the future of nanotechnology, "There's plenty of room at the bottom."

From the point when a computer can emulate an entire human brain, thirty doublings will bring us to over a billion brains,[5] and three doublings beyond that takes us to over eight and a half billion. Whether these doublings occur every one, two, or three years, this future is rapidly approaching. As we will see in the next chapter, there are several reasons to expect that the rate of doubling will actually increase in the future.

But will digital emulations truly be able to replicate the human brain, mind, and personality? Won't something essential be lost in the process? Perhaps. It really depends on how accurate the emulation is. Inventor and futurist Ray Kurzweil gives the example of scanning a single neuron in the brain of a living person and replacing it with a digital chip that perfectly performs all of the functions of the original. Given this scenario, few people would question that this was the same person as it had been prior to the substitution. Nor would that change if another of that person's 86 billion neurons were replaced. If we do the same thing again and again, at what point does that person's humanity depart? At what point are they no longer the person they once were? So long as the artificial neuron is a sufficiently accurate reproduction of the original, that moment should never come. But how can this be?

The answer may lie in a tale written down nearly two thousand years ago by the Greek historian and essayist Plutarch, in about 75 AD. The "Ship of Theseus" belonged to the mythical king and founder of Athens, and as such was preserved through the ages. Great care was taken to keep it exactly as it was, and as an oar, or plank, or mast decayed, it would be replaced in order to preserve the ship for posterity. Eventually, every single piece of the ship had been replaced, in which case how could it be the same ship that Theseus had once sailed on?

The paradoxical answer tells us that it is and it is not. Functionally, the ship remains as it has always been. The paradox arises because no single part of the original remains. The Greek philosopher Heraclitus[6] suggested that the answer lay in the metaphor of a river. Heraclitus offered that while the water in a river is continuously changing, it still remains the same river. In essence, the conditions from which its state of "riverness" emerges are greater than any of its constituent parts.

We can personally be thankful for this paradox. Cells in our body are continuously changing, dying, and being replaced on a regular basis. Epithelial cells that form the blood vessels are replaced approximately every five days. Skin cells live about 2 to 3 weeks. Red blood cells have a lifetime of about four months. Fat cells can last for eight years. An often quoted, somewhat misleading statistic is that on average every cell in our bodies is replaced every seven years. With one exception: our brain cells are with us throughout our entire lives, never being replaced, even when they die.

Except that this is not entirely true. Many cells in the brain are routinely replaced, notably glial cells and neurons in the dentate gyrus, a part of the hippocampus.[7] But many other neurons, including cortical columns, apparently are never replaced. Which must therefore account for the continuity of our sense of self, our belief that we are the same individual we have always been. This is a reassuring notion, but again, there is one significant problem. Every atom in our body, including those in the neurons, is replaced approximately every five years. Which brings us back to our original paradox.

Apparently, there is something more to our sense of self than the mere permanence of the individual components of our brain. If every atom is changed, then every cell is effectively changed as well. Nonetheless, our minds remain. Therefore, equivalent processes and structures must yield equivalent properties, at least within the ability of ourselves and those around us to note any distinguishable difference.

Building from that observation, we can assume that given sufficiently accurate reproductions of our neurons and their weightings, as well as other structures, our minds will remain much as they were, even if those neurons are artificial.

This then raises a number of ethical and legal issues. If you have your neurons completely replaced with artificial ones and still feel you are the same person you've always been, does that truly make you the same person in the eyes of the law? Assuming the answer is yes, what does that mean for your cognitive clones who are also emulated, except that in their case or cases, the emulation has been done using software?

If all of those same neural processes can be modeled identically inside the computer, then based on the preceding arguments, how can the clones not be equivalent? And if they are equivalent, shouldn't the clones also be considered the same person by our legal system?

Cognitive clone technology would create some unique conundrums for the law. If a virtual double commits a crime, is the physical host also guilty of that crime? What about any other doubles that were instantiated at the same time? If you conclude that they are not and that only the perpetrating version of that person should be punished, then what happens after that cognitive clone has reintegrated with its host? Has the criminal been eradicated? Or now that they are part of the memories, experience, and existence of the original person, do we have to punish the physical person too? Also, how do you punish a clone? What is a suitable punishment for premeditated murder by a conscious but virtual being?

You can see how this sort of technology is going to turn much of our thinking upside down. For instance, how do inheritance law and taxation apply if you leave your property to your cognitive clone? In a manner of speaking, aren't you leaving the money to yourself? Or are you? Without a great deal of legal forethought, wealth is probably going to become concentrated in ever fewer hands. Very possibly virtual hands.

From the standpoint of personal experience, uploading raises many other issues as well. While uploading one's mind to a computer could theoretically allow a person to live indefinitely, it would be a matter of perspective. For other people, family, friends, and acquaintances, the uploaded version of that person might truly be indistinguishable from their previous self. The uploaded double too may feel they are the same person they've always been, as they continue their once physical life, only now in digital form.

But for the original physical person, life would be very different indeed—because it simply wouldn't exist. Though the double may live on indefinitely, that doesn't affect the experience of the original. Though the physical brain may have been scanned in detail and replicated amid the bytes and registers of some immense supercomputer, the source person will never experience this extended life. That person will age, grow infirm, and die, unless his or her own brain and body are somehow physically augmented to live on through the ages. The peace, terror, or indifference of death will still be the last thing that person knows.

Almost ironically, the double will likely go through a similar experience over long enough timescales. For while we may wish that the digital versions of our lives live on forever, that isn't possible either.

In a 2017 paper, two mathematicians, Paul Nelson and Joanna Masel, showed that biological immortality is mathematically impossible due to intracellular competition.[8] Subsequent discussions with them confirmed that this limitation would extend to software and digital replications as well. The act of copying invariably introduces errors, as do competitive elements, such as cancer in biological systems and software viruses in electronic ones. The problem lies in the fact that if you have a highly complex, low-entropy system, there's nowhere to go but down. Whether relying on an immune system or intelligent nanobots or some form of software error checking, there will be either false negatives or false positives, because no decision-making process is perfect. Increase the system's vigilance, and you increase the number of false positives, killing off too many of the healthy cells. In a biological system, we call this an autoimmune disorder. Because the system cannot perfectly catch and kill all malignancies, whether cancer or malware, there will be false negatives, too, which invariably leads to degradation and aging. In the end, it's the age-old war against entropy all over again. Unless we can maintain perfect error checking—something that is effectively impossible over long enough stretches of time—the quest for true immortality remains quixotic.

It's important to note that just as the development of global society has accelerated our appetite for energy, the pursuit of augmented human intelligence not only perpetuates but augments that appetite. Even if computing and AI eventually attain human levels of cellular energy efficiency, which itself will be a huge challenge, we will not be developing this efficiency to merely maintain the status quo. Accelerated intelligence will require commensurate amounts of additional energy. Backups and other forms of redundancy will add to this, as will the extra error checking needed to reduce (though not eliminate) the degradation inherent in the system. In short, our future energy use is destined to climb a steep exponential curve for the rest of our existence as a species.

Finally, it is interesting to consider that in our pursuit of increased brainpower and our dream of immortality, we are essentially working to maximize our future freedom of action. In the course of doing so, by concentrating and using still more energy, we raise the energy rate density of our existence. Though this reduces entropy locally, it will increase it universally, accelerating the running down of the cosmos little by little. It is a pattern we will see again and again in the coming chapters, as we continue our journey to the end of the universe.

CHAPTER 16

A MULTIPLICITY OF MINDS

"What are humans for? I believe our first answer will be: Humans are for inventing new kinds of intelligences that biology could not evolve."
—Kevin Kelly, editor, author, techno-visionary

Archaeologist Prime pondered the residue of data fragments that remained on the ancient quantum storage units. After a long pause of several hundred milliseconds, the renowned historian turned to his trusted assistant, who perched beside him.

"Excellent work, AI-315Z22. This was a challenging bit of forensics, but you managed to retrieve far more than I would have thought possible. This data artifact fills a very important gap in the digital fossil record."

The diminutive AI trilled its pleasure at the compliment.

"Yes," Archaeologist Prime continued. "This provides solid support for my thesis that the second Cambrian Explosion—the vast proliferation of intelligences during the twenty-second century—was a result of the AI-directed architecture changes that were made in that century's first decade. After that, the many intelligences that were created by humans and other AIs expanded very rapidly. The Cognitors split off from the Sequellians, which led to the entire line of Parallelists. In a fraction of a millennium every major phylum of intelligence we know today had been generated. You're a direct descendant of that era's earliest forensibots."

The AI warbled approvingly.

"Of course," the archaeologist continued, "those bots were so very primitive by our modern standards."

A series of high-frequency waveforms emanated from the assistant AI.

"Don't take the tone with me," Archaeologist Prime snapped. "I wasn't casting aspersions on your progenitors. I was merely observing how far you've come. How far we've all come. For instance, I wouldn't be the person I am today without my unique combination of wetware and hardware. Each of our *intellineages* originated and emerged from that era of rapid speciation that followed the rise of *Homo hybridus*. The merging of human and machine. From there began the evolution of the vast ecosystem of intelligences we know today. Including the specimen you just unearthed."

AI-315Z22 responded with an elaborate trill.

"You're welcome."

For three and a half million years, *Homo sapiens* and its hominid ancestors have occupied a unique niche here on Earth. Beginning as a relatively unexceptional primate, we began to develop, through a series of fortunate circumstances, an array of unique and beneficial cognitive functions. Regions of the brain involved in cognition developed and specialized for different purposes, most of which were essential to the perpetuation of our species.

The forming of edged stone tools not only helped us survive; over time it enhanced our motor control, attention, and ability to learn and recall complex sequential steps. The ability to transmit that knowledge to successive generations presumably enhanced the regions of the brain essential to communication and social interaction. The ability to create stone tools and to pass along the knowledge to do so, along with some fortuitous genetic changes, were later crucial to the development of complex syntactic language. These and many other factors resulted in our becoming the only advanced technology-wielding species on this planet. It has remained a very exclusive club.

However, our human perspective has tended to bias our ideas about what intelligence is, how it functions, and what its purpose might be. When we discuss the future development of advanced AIs, we often refer to them as artificial general intelligences or AGIs. But there is no such thing as a true general intelligence. The many intelligent agents that make up the human mind are far from general but rather are highly specific to our species, and the same could probably be said of any other "general intelligence" in the future. We refer to what goes on in our minds as general intelligence for no better reason than this is the only form of mind we truly know. It's a very anthropocentric perspective. As Nagel's bat reminds us, we are unable to truly comprehend the mind of another species, and this will be true of artificial intelligence as well. But this does not mean other forms of intelligence are not valid. More importantly, there is no superior species of intelligence. Given our different origins, biologies, body plans, and ecological circumstances, the best intelligence is the one that perpetuates your own species in its particular niche.

Perhaps we need something that captures what we mean by general intelligence in a less anthropomorphic sense. According to professor of machine learning Thomas Dietterich, "'General intelligence' is a system that can behave intelligently across a wide range of goals and environments."[1] In many respects, this seems to embrace much of what we mean when we use the term.

The ability to introspect and to consider our place in the world also inclines us to assume we are more special than we are. Much of this ability directly results from our acquisition and development of language. Without the vast number of concepts and relationships this makes possible, not only would our various forms of metacognition be restricted, but our ability to share that knowledge with each other would be as well.

This ability to originate and manipulate concepts has brought us to the point that we are now building technologies that span a range of different potential intelligences. Though none of these is yet what we would describe as general intelligence, many are superhuman in some very narrow range of what they do. Of course, this is only the beginning.

As we have seen, while AIs are capable of many phenomenal feats, they do not achieve them using the same methods that human beings do. This is a good thing, because many of these AIs' amazing capabilities would not be possible if we had somehow managed to configure them to approach the task as we do. The reason these technologies are so useful to us is exactly because they are so very different from us. A human being can't locate and access a specific sentence buried somewhere in the depths of a book or document from among all the millions of writings on the planet, but search engines do it repeatedly, continuously, every day, in fractions of a second. Likewise for complex calculations that might take years to perform by hand. Or for identifying hidden patterns mined from millions of documents or images. It is the ways these systems are different from us, from the way we think, that make them so valuable to us.

Certainly, the AIs we routinely deal with are narrow and as a result brittle, but that will soon change, probably far more rapidly than most of us expect. But how adaptable can these AIs become, and what is the cost of imbuing them with a broader, more general intelligence?

Whether we are discussing nature or technological design, specialization comes at a price, as does generalization. This goes for intelligence as much as it does for any physical feature. Every animal species has a finite amount of energy to dedicate to its specific phenotypic features. For instance, a caribou expends a large portion of its available energy growing and maintaining its massive antlers. Similarly, our brains use about a quarter of all of the energy we consume. Each of these limits the physical resources available for developing and maintaining other traits and functions. Similarly, a computer that is optimized for calculating planetary orbits is not going to excel at natural language processing. It might be possible for it to be designed to do both, but it will perform neither as well as a dedicated system would. A machine that can fly like a bird, swim like a fish, and dig like a mole will not do any of these particularly well, certainly not so well as three separate devices optimized for each purpose.

For the time being, these developments will continue to be driven by market forces—human-centric market forces, to be exact. The nature of the competitive forces that drive corporate strategy and decisions, individual accomplishment, and national agendas, along with our innate curiosity, will remain the only motivators for at least several more decades. These market forces will continue to drive specialized AIs, because we need these systems to navigate and control our accelerating, increasingly complex world. Anyway, we already have the ability to create human-equivalent "general" intelligences; they are called *babies*. The value of AIs to us is their ability to do things we never could, whether it is their speed of processing or their ability to pore over vast amounts of data to identify otherwise unseen patterns. Trying to give all of these "specialists" general intelligence would probably mean they wouldn't be able to perform their specialized tasks anywhere near as well as they would otherwise.

This isn't to say AGIs won't be built, just that it will be at a cost to their other more specialized functions. However, it may still be possible for an AGI to avail itself of other systems on an as-needed basis, just as we do. Whether it does or not will continue to be determined by basic economics.

In time there will probably come to be many new players in town. At the point that AI becomes capable of enough volition and ability, we may begin to see it design new AIs as well as alter its own architecture. As we have seen with algorithmic-based design in the past, this can yield highly unusual and unexpected results.

A classic example is an evolved antenna created by NASA's Ames Research Center in 2006. Working with genetic algorithms, researchers used a computer to design an antenna for NASA's Space Technology 5 (ST5) mission. The resulting design was unlike any a human engineer would ever come up with. The antenna, which has often been likened to an elaborately bent paper clip, was developed using an algorithm that emulated several aspects of natural selection. Multiple candidate designs were introduced into the program, then moderately altered using a combination of genetic operators, recombination, and mutations that iteratively introduced changes. These were then passed to a "fitness function" that determined which designs were the most successful so that they could be passed on to the next generation. The entire process repeated hundreds of times until the desired specification was met.

Similarly, as computers and AIs have become increasingly involved in their own design and redesign, we should probably expect that they will discover approaches that would otherwise elude human computer architects. From new methods for optimizing novel functions to innovations based on bio-inspired models, we can expect to see computers generating approaches and designs unlike any the human mind would ever conceive.

So here we have a number of eventual drivers of the development of multiple intelligences: market forces, be they corporate, political, or individually motivated; decisions made and executed by sufficiently capable AIs; and finally, human curiosity, which will continue to lead some people to explore new configurations of their own minds, once the option exists, just as people and cultures have used drugs to explore alternate realities for millennia.

All of these drivers could one day contribute to a vast ecosystem of intelligences that will come to occupy our world and beyond. Just as with biological species, these intelligences could eventually fill the many niches that are waiting to be exploited. As new intelligences develop, they will in turn lead to still more niches, just as new plant and animal species generate new niches for other species in nature. In the end, the proliferation of different intelligences and minds will come to span an immense domain of which human biological intelligence may only occupy a very small region.

This is a good point to talk about intelligence, brains, and minds. Each of these concepts has many definitions, but for the sake of this discussion, let's explore a few of the differences.

From the beginning of this book, we have looked at the idea that intelligence in its broadest sense is about a system's ability to adapt to change in order to maximize its future options, particularly its survival and perpetuation as an individual and a species. This definition can apply to a single-celled microorganism, a squid, or a human being. As AIs grow in complexity, it could also come to be a function of their processing as well.

At the time of this writing, a brain remains an organ of an exclusively biological nature, grown in a manner in which a series of recurring structures are organized according to instructions defined by DNA. While we may colloquially refer to certain artificial intelligences as brains, none have yet come close to the complexity and capability of a mammalian brain. Even the full emulation of a fruit fly's brain, with its mere 135,000 neurons, remains elusive. While we may refer to AI programs and later AGIs as brains, this usage is currently just a metaphor, an effort to conceptualize what we are attempting to build.

Mind, on the other hand, is an emergent property of brains or potentially other structures that enable intelligence. Call it a subset of intelligence, if you will, because certainly not all forms of intelligence result in mind. The brain of a roundworm effects intelligent behavior but probably does not result in any emergence that we might call a mind. There is probably a minimum threshold of complexity that must occur for anything like mind to emerge. Will AI ever attain this level of complexity? Very probably. How many types of intelligence and mind will eventually come to occupy this ecosystem? That's a far more difficult and probably unknowable question. Nevertheless, it's worth speculating on the many forms they could eventually take.

There are many different ways we can look at the future of potential intelligences. One way would be to consider the various methods for building, growing, or otherwise creating them. Currently, the biological path that has been driven by millions of years of evolution has had the greatest success. Are we nearing the limits of what can be achieved by the interconnecting of neurons, or are there still some gains waiting to be made?

Computer architectures built from silicon and gallium arsenide chips and processing units, as well as future substrates, suggest a continuing path for growth. Engineered processors have some immense advantages over biological cells but have yet to exhibit the level of complexity found in animal neural structures. Is it simply a matter of time and knowledge before computers and AI become capable of realizing similar results? Or will the differences between biological and technological substrates prevent AI from ever achieving truly advanced intelligence?

Beyond silicon, there are numerous potential substrates and strategies that, while related to current-day computing, deserve their own mention. Quantum computing is still very much in its infancy, but the ability of qubits, the basic building block of quantum computation, to exist in superposition suggest many interesting possibilities. Quantum superposition, the state of qubits being neither off nor on, unlike digital bits in a binary computer, allows for different and unique approaches to optimization problems. These problems take many forms. A classic example is the Traveling Salesman Problem, in which the shortest route through a series of cities is calculated. As the number of cities climbs, the calculation soon becomes incalculable for classic computers, potentially requiring more time to solve than remains in the life of the universe. Yet a quantum computer could quickly find the solution. Whether or not quantum computing could eventually be used to build an actual intelligence, however, remains to be seen.

Synthetic biology is another field that could one day allow us to synthesize either life-form or organs that will function independently or be used to enhance existing biological neural functions. On the one hand, we still have far to go in our development and understanding of the processes involved. On the other, we may be able to accelerate the process by availing ourselves of mechanisms already long established by nature.

Molecular nanotechnology involves the direct manipulation of atoms and molecules in order to manufacture machines at the atomic level. The field is still in its infancy and must contend with numerous challenges, particularly interference from various quantum effects. But once these challenges are overcome, assuming they can be, all sorts of new strategies and materials will become possible.

There are no doubt numerous other platforms on which we and future intelligences will one day build still other forms of intelligence. Most potential architectures likely have yet to be discovered, but they will be over time, as the benefits of accumulated knowledge compound.

Technology visionary Kevin Kelly has written about the idea of a taxonomy of minds,[2] the many possible ways intelligence may eventually manifest. He focuses mostly on function, rather than the substrate on which any given mind is based. While Kelly admits that his list of possible classes of minds is far from comprehensive, it is nevertheless a long one. Which of these will humanity be able to share in and which will remain part of an exclusive club to which we will never be admitted?

The easiest of these classes to imagine, according to Kelly, is one that is like the human mind, only faster. I would add that a faster human mind is probably the only mind we are capable of truly imagining or comprehending, because for all intents and purposes we are talking about ourselves. Nearly every other form of intelligence that follows is subject to Nagel's arguments about the incomprehensibility of different minds and forms of consciousness.

Perhaps the category of minds we know best from science fiction is the aggregates. These include a global supermind composed of millions of individual dumb minds in concert, perhaps akin to a colony of ants. Hive minds, which are made of many very smart minds that are unaware they are part of a hive, might be another form of aggregate. As is a "Borg mind," made up of smart minds that are very aware of the unity they form.

Another category includes the makers of minds, those minds capable of imagining and/or developing other minds or intelligences

greater than their own. Some in this category may be able to imagine the feat of building another mind but are incapable of actually
realizing their vision. Will this be our lot? Or will we be our planet's,
perhaps even the universe's, means of bootstrapping all of the future
intelligences to come?

Other types of minds in this category include those that are on
the threshold, capable of creating another mind only once, whether
intentionally or by accident. Other minds more capable in this capacity might be able to create greater minds many times in many different ways. Then there are those that manage to create greater minds,
that themselves are able to create still greater minds, and so on. How
long would any of these steps take? It really depends on the method,
substrate, and intent of the minds involved.

Another possibility is a mind that can change itself at one or more
fundamental levels, altering its own processes. While at first glance
this possibility seems like something exclusive to computing and AI, it
could well be within our capabilities too, if a suitably powerful human
mind augmented by nanotechnology ever exists. As dangerous as this
kind of alteration may sound, let us not forget that for many millennia people have done something similar with their use of natural and
synthetic mind-altering drugs. Certainly, a suitably advanced mind
would sandbox[3] and extensively test any potential alterations before
modifying its basic operating code.

Other minds may benefit in different ways from the assistance of
other intelligences. A mind that takes a long time to develop, requiring a protector mind until it matures, would bear certain similarities
to human babies and children that are unable to fend for themselves
living under the protection of older humans for many years.

A related possibility that might advance human minds would be a
mind trained and dedicated to enhancing another intelligence, such as
ourselves. The twist here is that it might be so customized to the user
that it would be useless to anyone else.

Cloneable minds occupy another sector of this domain. Kelly has
postulated a mind capable of cloning itself exactly, many times over,

much as a computer copies a file. Another type of mind might be able to remain connected to its many clones. Finally, another cloning method might allow a mind to repeatedly migrate from platform to platform, effectively achieving immortality. As noted in the previous chapter, immortality may not be entirely achievable due to the realities of entropy and thermodynamics, but certainly such an approach would have the potential to extend the life span of such a mind considerably.

Other possibilities include a nano-mind that is the smallest possible self-aware mind, presumably a product of nanotechnology. Or perhaps a mind that never erases or forgets anything. Again, the fundamental nature of entropy may place limits on this "never."

An anticipator mind might specialize in scenario- and prediction-making. As discussed earlier, the power of predictive analytics already suggests that such an ability is not out of the question and, if sufficiently powerful, might even have us questioning the concept of our own free will. After all, at what stage do we admit that if everything is predictable and therefore predetermined, then perhaps we are not the free agents we imagine ourselves to be?

Then there are the minds based on quantum computing, which will almost certainly be unfathomable to us. As noted earlier, such a mind may be a possibility if we can ever gain enough control over the properties of matter at this extremely fundamental level. While some theories have suggested that the human mind somehow exploits quantum behaviors in order to achieve consciousness, these theories have been widely refuted, not least because of the inhospitableness of our warm, wet brains to using such phenomena in a controlled way.[4]

But the truly interesting minds, I think, will be the hybrids, which could be countless. Even in this far from exhaustive list, there is such an enormous number of possible combinations that it would boggle our twenty-first-century minds. As we find possible ways to combine different substrates, that number would climb even higher.

Much has been written in science fiction about *uplifting*, different ways of raising the intellectual capacity of other animal species.

While this may stay in the realm of fiction for a few more decades, we shouldn't discount the idea entirely. A range of different research suggests that such uplifting may one day be within the realm of possibility.

In 2014, a team led by Ann Graybiel of MIT introduced the human form of the forkhead box P2 (FOXP2) gene into mice in order to learn its effect on learning and neural plasticity.[5] This gene has been closely associated with the acquisition of language in humans beginning some 200,000 years ago. Not surprisingly, the study did not result in talking mice. Instead, the altered subjects were able to more easily navigate a maze and form memories because the gene facilitated transitions between declarative and procedural learning. This ability could have also adapted the human brain for speech and language acquisition.

In another study, a team of researchers from Wake Forest Baptist Medical Center, University of Kentucky, and University of Southern California tested a neural prosthesis on rhesus monkeys that had been trained to select images in what is called a delayed match-to-sample task. The monkeys were trained until they could consistently perform the task with 70 to 75 percent proficiency.[6] Using a ceramic multi-electrode array, the researchers recorded brain activity within prefrontal cortical mini-columns of neurons in the L2/3 and L5 cortical layers. They then suppressed communication between those layers by administering cocaine, which alters dopamine reuptake in the monkeys' brains. Implementing a multi-input multi-output (MIMO) nonlinear mathematical model, the neural prosthesis was then used to stimulate the neurons—simulating and replacing the firing patterns that would have normally occurred across those layers of the brain, circumventing the suppressed communication. Even more interesting, when the same prosthesis and MIMO model was applied to other trained monkeys under normal conditions, their performance actually improved beyond their control 75 percent proficiency.

To be clear, this experiment was not performed with the intention of uplifting rhesus monkeys but rather was an early animal-stage test of methods these researchers would later apply in human testing. In

2018, the team reported on their subsequent human studies in epilepsy patients, in which they used a similar MIMO nonlinear model to mimic the subjects' neural firings, improving episodic memory by 35 to 37 percent. This is the most common form of memory loss affecting Alzheimer's, stroke, and head injury patients. The prosthesis will hopefully be used one day to replace this sort of lost function in patients.

In a somewhat more unusual experiment, human glial progenitor cells (GPCs) were engrafted into neonatal immunodeficient mice.[7] These astrocytes, a subtype of glial cell, are twenty times larger in humans than in mice and far more complex. Nevertheless, they proliferated in the chimeric mice as they matured. (In genetics, *chimeric* refers to the fusing of cells or DNA from two or more distinctly different genotypes or species.) Upon testing, long-term potentiation (LTP) of neurons was found to be significantly enhanced, as was the mice's ability to learn a range of tests, including maze navigation and object-location memory tests. Similar engrafting using mouse GPCs yielded no such enhancements, indicating the improvement in learning and plasticity was due to the introduction of the human cells.

So here we have three very different tracks that could one day be used to uplift animal intelligence: genetic modification, neural prosthetics, and genetic chimerism, in which cells from one species are combined with another. As evidence mounts that such uplifting may actually be possible, the question becomes less about *if* we can do it and more about whether we *should*. There are numerous people in both camps: those who believe we owe it to the animals if this is within our ability, and those who see it as a sort of cognitive imperialism in which we impose greater intelligence on species that is unasked for. It is certain to remain an ethical dilemma for decades, if not centuries to come.

Of course, all of this leads back to the enhancement of human intelligence and the many ways we may one day augment ourselves. So many of the forms of intelligence and mind that eventually come to exist in this developing *intelliverse* may eventually descend from ourselves. But to what end? Will all of this lift us up or will it diminish us?

The future of the human mind is a very open question, not least because this is the first time in our history—in the history of our planet, for that matter—that we or any other intelligence has had a say in the matter. Until now, intelligence and its emergent properties of mind and consciousness have been exclusively a product of evolution. There has been no guiding hand but rather a set of randomly determined processes and deterministic selection. But now we are entering an entirely new stage of our development in which we will have influence over the paths our species will follow in the future this will bring.

Why would we do this? Why will we not leave well enough alone? Because we are human beings, those ever-curious apes who are continually exploring the edge of the known and unknown, the accepted and unacceptable.

Much of how we handle this will probably be a matter of time and perspective. What is outrageous or unfathomable in one era can completely transform within only a generation or two. As we look ahead to the end of the century and beyond, what could this mean for humanity?

Certainly, one significant possibility would be that humans would no longer be the only advanced technology-wielding intelligences or minds on this planet. We would have become part of a much larger, perhaps even vast community. Some intelligences we would no doubt interact and work with, but there could easily be those we would not. Though every AI we know today is one we have designed to work with us or for us, there may be other tracks that generate new intelligences without any input from us. How many different kinds of minds might come to exist? Beyond the limited taxonomy described earlier, consider just the many forms of human intelligence. If human psychometric tests are anything to go by, we could see developments in numerous different directions. For instance, according to Howard Gardner, whose theory of multiple intelligences first became popular in the 1980s, people are composed of differing proportions of eight different intelligences: visual–spatial, linguistic–verbal, interpersonal, intrapersonal, logical–mathematical, bodily–kinesthetic, musical–rhythmic, and naturalistic. Whether we accept these as distinct intelligences or simply as talents, the exact combination contributes to making each of us the unique person that we are.

Now consider all of these represented as multidimensional vectors that denote the direction and magnitude of each category. How many possibilities could exist? How many niches could these fill in the cognitive ecosystem? It is probably safe to say that the range of human minds would eventually come to represent a very small region in the

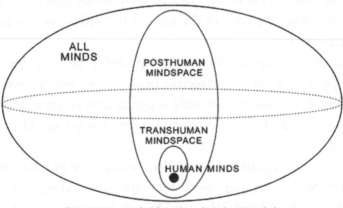

Potential space of all human and nonhuman minds

Based on a graphic by Eliezer Yudkowsky

midst of a far larger domain space of transhuman minds, posthuman minds, and, finally, all other possible intelligences and minds.

That would be only the beginning. With human intelligences, we are limited in type and level by our biology. But as new nonhuman intelligences proliferate, they could potentially cover a much more extensive range than their human counterparts. Also, unlike in people, the range of these new intelligences would be capable of developing and changing rapidly, especially when compared with evolutionary and even natural cognitive timescales.

So, nonhuman intelligences could eventually cover a much larger range of intelligence factors. In fact, it is probable that we would one day see varieties of ability and perception in AIs and ASIs that we currently don't have a conceptual precedent for. For instance, an AI that has only intuitive skills, but absolutely no analytical ones. Or an intelligence that has impeccable social skills and can comprehend and navigate a six-dimensional Calabi-Yau manifold as described in string theory, yet is unable to communicate with us verbally. Or perhaps an empathy bot that feels everything you do, but magnified. Many of these may make no sense to us, and the exact reason some of them would ever exist we can only speculate on, but as new intelligences and therefore new niches arise, we may find there are needs for all of them. Once the door opens, all manner of possibilities will rush through to exploit new opportunities.

Many of these intelligences will be our aides and allies, but certainly not all of them. How will these intelligences and minds behave as they vie for energy, data, and influence? What strategies will evolve in order to survive in this world of cognitive competitors?

According to *cognitive niche theory*, hominids were able to evolve and use their ability to reason and plan in order to occupy a unique niche in the ecosystem. Our ability to use cognition to respond to complex situations, much more rapidly than is possible for evolved traits, provided us with a singular advantage. This niche had the added advantage that once we filled it, we presented little opportunity for any competing species to get a foothold that might allow it to benefit by a similar strategy. As a result, the first species to dominate this niche—ourselves—stood alone because no other species ever had a fighting chance.

What happens when we are no longer the only intelligences that can reason, plan, and abstract? What happens when we're no longer the best at what we do? How does this transform the ecosystem and our future?

The reason we were able to excel was not just that our cognitive processes were inherently superior in terms of the options they provided. It was that they were many orders of magnitude *faster* at responding to new challenges. As individuals, as communities, and as a species, we were able to "fail fast" and learn from the process. Now, if we are suddenly joined by an array of intelligences that can fail, learn, and adapt faster than ourselves, again perhaps by many orders of magnitude, where does that leave us? The traditional tooth and claw approach of natural evolution holds as little appeal as it does chances of success. Possibly our best response will be to use our heads in order to establish what are essentially new pacts and partnerships.

As this battleground unfolds, what alliances and forms of cooperation will be made? If the best way to deal with a powerful antagonist is to join forces with a fellow intelligence, why wouldn't that be our preferred solution? Taking this a step further, would we one day choose to combine human intelligence with other forms, leading to human-machine symbiotes, cognitive collectives, and even hive minds? (To be clear, I am not welcoming or advocating this, but anticipating the possible outcome of a projected trend.)

This scenario feels very foreign to us, not least because we have always been relatively solitary minds. The closest we ever come to anything beyond this is through social interaction, the application of our collective intelligence. From idle conversation to deeply empathic bonding is when we come closest to merging with other minds. Advocates of meditation and certain forms of spirituality also claim a sense of being part of something larger than themselves, though how much of this is an artifact of our own cognitive peculiarities has yet to be determined.

But this is all rudimentary compared to the potential aggregating of minds. In much the same way we have seen new emergences arise from the aggregating of cells, insects, and neurons, what might a true network of minds lead to?[8]

What would it be like to actually join with another mind? We've already discussed the idea of sharing experiences and thoughts using brain computer interfaces. What other ways might this manifest? If our brains are an amalgamation of different modules and processes that have evolved and integrated over time, then each is, in a sense, its own rudimentary intelligence. Nothing that could lead to mind or consciousness itself, but when combined with all of our other modules, it yields something greater than the sum of its parts. Now, if we replace one of those modules with a technological prosthesis for whatever reason, be it therapeutic or with the intent of augmentation, we will have integrated with a foreign intelligence. Initially, it will be intended to be indistinguishable from its biological predecessor, but over time, as that prosthesis becomes more capable and is altered from its original purpose, the person's mind is going to change.

Will this change be for the better? From our current-day perspective, we may not think so. But the technological choices we make and have made, whether in the nineteenth century, the twentieth, or the twenty-first, are not based on the past. We make choices based on the present and our anticipation of the future. Of course, humanity does not always make the right choices, but overall, we trend toward our best interests and our survival as a species.

Whether these developments occur a hundred or a thousand years from now isn't really important (though I'm betting on one to two hundred years). What matters is whether or not this path allows our species to survive and ideally to proliferate as some version of the human race. Obviously, all of this enhanced and augmented intelligence is for nothing if we abandon our humanity and all we are. But we also must remember that change is part of life. Our species exists because earlier species such as *Homo heidelbergensis* and *Homo habilis* evolved into new species, species that were better adapted to their times and conditions. We don't feel diminished by the loss of our *Homo heidelbergensis*-ness.

As we continue on our path into the future, we will likely find there are ways we could better adapt to the challenges it brings. While in the past we have evolved at a pace sufficient to deal with our changing environment, change comes so much more quickly now, far more rapidly than natural selection can adapt us. The need to be able to deal with the exponential growth of data, to be able to distinguish fake news from real, to parcel out our attention in an appropriate and managed way, are but a few examples of how our world is vastly different from that of our ancestors. All of these and more can be better managed with the tools we will build. But over time, we are likely to find these tools are better and easier to use if they can more closely integrate with our senses. In the end, we may find that our devices, our tools, our digital personal assistants work best when they are part of us.

Of course, many of our future changes won't just be cognitive; they will be physical as well. This will be especially important as the environment we live in transforms, whether this is in response to climate change or to some other environmental disaster. But it will be especially needed as we finally make the big jump and begin to colonize space, as we will talk about next.

CHAPTER 17

AWAKENING THE UNIVERSE

"If we want to continue beyond the next hundred years, our future is in space."

—Stephen Hawking, theoretical physicist, cosmologist, and author

---

The year is 2265, and our ship is parked in orbit around the fourth planet in the Tau Ceti system. Located 11.9 light-years from Earth, the rocky Tau Ceti *e* is about four times larger than our own planet, occupying an orbit half as far from its sun. Its metal-rich composition and location in the habitable "Goldilocks" zone of this solar analogue[1] makes it an ideal candidate for exploration.

We watch as a tiny probe enters the system, our timetrack-linked monitors set to observe it in accelerated time, much like watching a sped-up movie. The probe deploys a solar sail that rapidly decelerates it from the 12 percent of light speed maximum it has been traveling for more than a century. Yet, even at this speed, the relativistic effect on the probe's mass has been negligible.[2]

The probe deploys a set of smaller daughter probes that assume orbits around each of the star system's seven more massive planets, as an eighth probe enters the fourth planet's atmosphere and promptly sets down in a secluded region of its southern hemisphere. Each probe begins gathering data, which they regularly transmit to the mother probe for beaming back to Earth, a transmission that will travel for nearly twelve years before it is received.

Turning our monitors to the single moon of Tau Ceti *e*, we watch as one final probe descends to its surface. Before long, the probe briefly opens a hatch from which it dispenses a fine mist that quickly settles over the ground. Soon after, the regolith, the dull gray soil of this moon, starts to *churn*.

Slowly at first, then with increasing speed, the moonscape in that region begins to change. The image almost strobes as day becomes night becomes day, again and again and again. With each passing cycle, mounds of different color and texture grow up from the ground, almost as if they were eroding backward.

But they *are* growing, as a vast army of semi-intelligent nanobots mines the moon's surface, extracting and refining usable elements as no human miner ever could. Then, as though a switch had been flipped, the mounds stop growing and slowly begin shrinking as several somewhat more defined structures begin to coalesce in their midst. Expanding upward and outward, the structures resemble nothing so much as a set of gray icicles melting in reverse—that is, until they nearly reach their final form. Then it suddenly becomes apparent what they are: dozens of new probes that are identical to their mothership. Molecule by molecule, the nanobots have additively built a new generation of spaceships modeled after the one that delivered them here.

As mining continues in different regions of the moon, propellant and other fuels are manufactured and transported to the waiting probes. Finally, all is ready and the probes lift off the surface of the moon as one, overcoming its low gravity with ease. Then, as though of a single mind, they accelerate outward in multiple directions. Unfurling their solar sails, the probes slowly accelerate out of the solar system, putting first the moon, then the planet, and eventually the entire solar system behind them. They are off to repeat the entire process in as many other star systems, so that it can all be done again and again.

As for the sister probes they have left behind to seek out life and perhaps even intelligence on Tau Ceti's seven worlds during all of this: What did they find? The people of Earth will know once the mothership's signal reaches them—in just a little less than a dozen years.

For more and more of humanity, the call of the cosmos, the dream of other worlds, is growing. No longer just a romantic notion, the move into space is rapidly becoming imperative. We have come so far in these three million years that progress has left us perhaps more vulnerable than ever before. As famed physicist and cosmologist Stephen Hawking observed during a BBC Reith Lecture in January 2016:

> Although the chance of a disaster to the planet Earth in a given year may be quite low, it adds up over time, and becomes a near certainty in the next thousand or ten thousand years. By that time, we should have spread out into space, and to the other stars, so a disaster on Earth would not mean the end of the human race. However, we will not establish self-sustaining colonies in space for at least the next hundred years, so we have to be very careful in this period.

As we pass through the twenty-first century, we face an array of threats unlike any we've ever known. While it is fair to say that the threat of global nuclear war during the second half of the twentieth century was comparable, it was perhaps also more containable. The complexity of building nuclear weapons meant that those players with nuclear capability remained a fairly exclusive club.

But so many of the emerging technologies of tomorrow are or will be ubiquitous. AI and cybercrime will be everywhere. DIY biotechnology could result in a biological disaster that originates in someone's garage or basement. Misuse of CRISPR/Cas9 could fundamentally alter our species to a degree from which we might never recover.

The vision of leaving our planet "to boldly go where no one has gone before" is a romantic one, appealing to our sense of adventure, discovery, and individualism. But it can also be much more than this. It can provide us with a backup.

Throughout the history of life on Earth, many species have been decimated by natural catastrophes, famine, disease, and other environmental disasters.[3] Even *Homo sapiens* have nearly been wiped out several times.[4] In many of these cases, it was the isolation of certain fortunate populations that allowed their respective species to recover.

But this only works if there is somewhere that remains quarantined from the affected regions. Virulent pathogens, global war, and extreme climate change could theoretically reach every corner of our planet. Even extraterrestrial threats such as a large asteroid strike have the potential to wipe out most life on Earth in mere hours. The decimation of the dinosaurs 65 million years ago attests to that. A recent discovery of a cache of fossils in what is now North Dakota even corresponds to the very hour that the cataclysmic event occurred.[5]

While we have reached a stage when we can anticipate and potentially even respond to some of these threats, their number and nature are escalating. Before this past century, all of the possibilities for our extinction were of a natural origin, but more and more, the threats are of an anthropogenic nature. In other words, we are well on the way to being the cause of our own demise.

But if we have established colonies beyond this one world—if the moon, other planets, even distant solar systems become home to humanity as well—then these vulnerabilities disappear, at least as a threat of species-wide extinction. Obviously, this is not something that will take place overnight. The knowledge and technology needed to accomplish this will take decades and centuries to develop. The colonies will need time to grow into settlements and, in time, societies.

But it will be necessary, because if we remain exclusively bound to this one world, our long-term odds are not good at all.

Assume the chances of a planetwide catastrophe are one tenth of one percent in a given year. We might say the chances sound negligible, yet we are not talking about winning the lottery but rather the death of everyone everywhere. Forever. Not only that, but everyone who might have one day been born in the future would never exist. Far more concerning is what happens when these odds are extrapolated over time. Over the course of a century, the chances of an existential disaster leap to nearly 10 percent, with odds reaching fifty-fifty in just seven hundred years—about the amount of time that has passed since the beginning of the European Renaissance. If we use the more pessimistic figure of 1 percent odds in a given year, that flip of a coin milestone is reached in only 69 years!

Of course, we don't know the exact level of risk in a given year, and the value is likely to change over time. But the 2008 Global Catastrophic Risk Conference at the University of Oxford estimated the odds of human extinction by 2100 at 19 percent. The Stern Review for the UK Treasury in 2006 offered the only moderately less pessimistic value of 10 percent. The point, though, isn't really about whose estimate is the more accurate, especially if there's no one around to see who was right.

This is not to say that we don't need to keep striving to perpetuate our species and protect our environment here on Earth. We absolutely must. But it would be negligent for us to assume we are certain to succeed and therefore that there is no need to build a backup plan. It will be a tragedy of unthinkable proportions if we were to lose everything through our inaction.

If we accept this course of action, this destiny of extending our civilization beyond our lone planet and expanding our intelligence into the cosmos, what then? We will no doubt want to ramp up our efforts of developing the needed technologies that allow us to explore space. It will be to our continued benefit to increase the intelligence of our systems so that they can better anticipate our needs and protect us from the many dangers of living off-world. Most importantly, such systems will need to perform many of the tasks that will be too hazardous for our still vulnerable bodies to undertake.

These increasingly intelligent space probes will be used to go forth and scout ahead, learning, measuring, and reporting back to us on what they find in the outer reaches of this solar system and, eventually, much farther away in other star systems. For a very long time, these probes might be contained to relatively local star systems, but eventually the entire galaxy could be explored and colonized.

How could we possibly explore our entire galaxy? After all, the distances that would need to be covered are unfathomably enormous. How enormous? If we rescaled the universe so that Earth was the size of a penny, the nearest star system, Alpha Centauri, would still be more than sixty-five thousand miles away! Even traveling as fast as light, the maximum speed limit of the universe, it would take more than four years just to reach the nearest of our neighbors. Based on our current technology, our fastest spacecraft would take some 80,000 years to travel that far. Existing methods of chemical propulsion simply are not able to overcome a number of limitations in order to travel such a distance and so are not viable options.

However, research is being done on several potential technologies that could eventually travel much faster than this. Solar sails, for instance, use the minute pressure of solar radiation emanating from a star such as our sun to accelerate to speeds comparable to the fastest methods of chemical rockets. As the materials and designs of solar sails improve, it may eventually be possible to achieve speeds approaching 1 to 10 percent that of light, which would allow us to journey to the nearest stars in anywhere from a century to only a few decades.

This is of course still far too long for people to travel there, but the laws of physics place very significant restrictions on what we can do. We are limited to traveling at subluminal speeds—that is, slower than the speed of light. Einstein's theory of special relativity outlines how light speed is restricted to 186,282 miles (300,000 kilometers) per second. At least as importantly, the theory limits this speed to massless particles, which is why the photons that make up light and other forms of electromagnetic radiation are able to achieve this. All matter in the universe—protons, atoms, spaceships, or human beings—are prevented from ever reaching this universal maximum limit by the fact that they have mass. But why should this be?

One peculiar implication of special relativity is that as matter moves, it increases in mass relative to the inertial frame of reference. Because energy and mass are equivalent in terms of relativity and since nothing can move faster than light, as an object increases in velocity, space, time, and mass must change relative to the frame of reference, otherwise the speed of light would be exceeded and causality would be broken.

This change in what is known as the object's relativistic mass is so negligible for most of the speeds in our everyday world that it is virtually undetectable and can be ignored. So, most of the time we use an object's resting mass, which is sufficiently accurate for most classical calculations. But if we accelerate an object sufficiently, its relativistic mass will start to climb. At 10 percent of the speed of light, an object's mass will have increased by only about half a percent, but at 90 percent of the speed of light, that same mass will have doubled. From there, if its velocity could somehow continue to be increased, that object's relativistic mass would climb precipitously. But as the object's velocity increases, the amount of energy needed to accelerate it further climbs by an even greater amount. As a result, even the smallest particle of matter would reach infinite mass if it could ever miraculously attain light speed. Since this would require infinite energy, even near-light speed will probably be impossible for any type of physical spacecraft for the foreseeable future.

While it is true that there are a number of inventions and theories being explored that might one day overcome these limitations, we probably shouldn't hold our breath. Some theoretical technologies offer the promise of sub-light-speed travel. The EmDrive, which is currently being tested at NASA's Advanced Propulsion Physics Laboratory, is one of these.[6] But as of yet, the EmDrive's minute thrust appears to be an artifact of measurement that hasn't yet been fully explained. The Alcubierre drive is another theoretical means of propulsion that could result in faster-than-light travel, if it is ever realized. Unfortunately, the Alcubierre drive depends on the existence of an energy density field less than that of the quantum vacuum, which as far as we know is the lowest energy state possible in a quantum mechanical system.[7] To achieve a field of even lower energy density would require matter with negative mass, something that has never been discovered. According to Einstein's theory of general relativity, anything having mass is unable to ever reach or exceed the speed of light, but negative mass would get around this by "warping space," allowing the object to effectively travel faster than light. While there is some disagreement as to whether negative mass would violate any laws of the universe, its existence remains highly speculative.

Because of the limitation provided by the speed of light, much of the speculation about methods of exceeding it has revolved around the warping of space or the creation of a wormhole that creates a shortcut from one part of the universe to another. These are predicated on the idea that the shape of space can be changed by large masses, gravitational waves, negative mass, or significant amounts of energy. Unfortunately, even if these should ever be something we can manipulate and control, the amount of energy required would probably dwarf that of our sun. While we may one day be able to harness such vast energies, we have a good many stages to pass through before we get there.

Obviously, transport and time are significant factors when considering how human intelligence, technological intelligence, or any other form of intelligence can ever propagate in the universe beyond its place of origin. Beginning with our own exploration technologies, any communication with command centers back on Earth is hindered by the vast distances of space. Even communication with a probe on Mars can take anywhere from three to twenty-two minutes, depending on where our two planets are in their relative orbits. Now just imagine the issues that need to be addressed when those round-trip instructions need more than eight years to be sent and received! Beyond Alpha Centauri, distances to other stars and other worlds rapidly grow by orders of magnitude, meaning it's unlikely we will ever be able to control those probes once they are well on their way. Consequently, a significant amount of programming and intelligence will need to be built into any future spacecraft, so they can adapt to the conditions they find and make decisions for themselves.

For the relative near term, it seems a certainty that humankind in its current form will be restricted to living in our own solar system and, if we're optimistic and extremely lucky, eventually a few of our closest neighboring stars. Given this restriction, we are left with making the best of what we have. Rather than pushing out farther and farther in search of a planet that is as Earthlike as possible, we will need to find ways to adapt ourselves to planetary environments that fall within a broader range of conditions. Temperature extremes, radiation levels, and atmospheric composition and pressure are all factors we can adapt to using existing technologies, if they are not too far outside a certain range. Even those criteria will no doubt expand over time as technologies are developed that can protect us from greater extremes. Suits, shields, and other methods would allow us to live in hostile environments, just as astronauts have done since the earliest days of space travel.

But of course, in the long run, this is no way to live and would no doubt lead to many unnecessary deaths due to accidents and equipment failure. No, if we are to live in such environments for the long-term, spanning many generations, we will need a very different strategy. One that allows us to adapt to what would otherwise be hostile and even fatal extremes.

Technologies that allow us to fundamentally modify our bodies and minds would be one possible approach. Genetic modification and other biotechnological and nanotechnological techniques could provide the means of protecting our DNA and making our bodies more resistant to moderate levels of radiation. An oxygen-poor atmosphere could be sufficient if our red blood cells were able to transport oxygen more efficiently, through genetic modification. Alternately, the eventual use of nanotechnology could provide other means of providing oxygen to our cells. Nanotechnology scientist Robert Freitas has proposed replacements for red corpuscles called respirocytes,[8] which would provide considerably more efficient means of transporting oxygen using nanomachines. These respirocytes could operate across a much broader range of conditions, allowing a person to swim underwater or breathe in an otherwise toxic environment for extended periods of time and eventually indefinitely.

Nanotechnology could also protect our body's cells and organs in other ways, from reinforcing cell walls to shielding and even repairing DNA from radiation damage. Of course, as more aspects of our bodies and brains come to be restorable or even augmented using robotics, neural prosthetics, and neuromorphic chips, we will find ourselves less biologically based and more technological in nature, effectively becoming cyborgs. This progression from biological to bio-technological and perhaps even entirely technological will go hand-in-hand with our continuing pursuit of ever greater intelligence and therefore computation. This evolution would have the added benefit of allowing us to become more resilient to the biologically harmful conditions of non-Terran environments. Many people will be concerned about such a metamorphosis, even if it is gradual, and rightly so, from our present perspective. But we may find in time that such transformations are the least unpleasant of our options as we work toward the long-term survival of our species.

Continuing with our colonization of space, there are of course other approaches. Terraforming is the engineering of a planet or moon's ecosystem at a global scale in order to make it more habitable. While this may be within our ability in a century or two, the scale and level of interconnected complexity in a planetary ecosystem will probably make this a far more challenging process than many of its proponents think. The time frames needed will also take us much farther into the future before success is realized. Just consider the challenges we currently face addressing Earth's own carbon dioxide levels and the resultant global climate change it is bringing about. Even smaller, self-contained systems such as Biosphere 2 and the International Space Station illustrate just a few of the significant challenges that efforts to direct and balance the engineering of an entire planet's ecosystem will face.

However, at some point, as technology advances, we may find we have new opportunities available to extend our reach throughout the galaxy. Self-replicating nanotechnology may be one of these. While this technology is still in the earliest stages of development, by the second half of this century, we will have become much more proficient with it. Not only will this allow us to build and control machines at hitherto unthinkably tiny scales, we will also use it to reverse engineer and improve on biological processes at the intracellular level. Additionally, self-replicating nanotechnology almost certainly will make possible space travel strategies that would previously have been impossible. For instance, a gram of carbon contains approximately 5×10^{22} or 50 billion trillion atoms. By comparison, a typical human cell weighs about one nanogram (10^{-9} g) and is made up of about 100 trillion atoms. Engineered nanobots could be many orders of magnitude smaller than one of those cells, allowing billions of nanobots to occupy a volume of only one cubic centimeter. Assuming a nanomachine that is a cubic micron in volume, a trillion could fit into such a minute payload.

But it isn't just their size that would make nanobots so powerful. Given their potential ability to self-replicate, repeatedly creating duplicates of themselves, the number of nanobots could increase exponentially, much as a colony of single-celled organisms does. Of course, nanobots could be programmed not just to replicate but to differentiate so that they could perform all sorts of different roles.

As we saw in the scenario at the beginning of this chapter, space probes might one day travel to other star systems using solar sails. A payload of nanobots would be of extremely low mass, allowing it to be rapidly accelerated using this method of travel, perhaps after moving beyond planetary gravity wells using other forms of propulsion.

Because solar radiation pressure follows the inverse square law, as the solar sail's distance from a star doubles, the pressure on its sail is reduced to only a quarter of what it was. In the case of our own solar system, there might not be enough energy for additional acceleration at about half a million miles out from the sun, somewhere near Jupiter's orbit. (This could improve with the use of new materials and would also differ from one star system to another.) One possible solution to this would be to locate a solar-powered laser at some midway point that could provide the needed boost to continue accelerating the solar sail out of the solar system, allowing it achieve a much higher cruising speed by the time it reaches interstellar space.

Plans for such a concept are already being explored through the Breakthrough Starshot program, which seeks to use four-meter-square solar sails to reach our nearest star by accelerating centimeter-sized vehicles (dubbed StarChips) using a one-square-kilometer laser array. The laser array would have a combined output of 100 gigawatts and might need to be space-based due to atmospheric interference. Theoretically, it would allow the tiny craft to reach an acceleration ten thousand times that of gravity during its initial ten minutes of flight. (Such an incredible acceleration precludes having any form of biological material on board, as does exposure to solar and cosmic radiation once beyond the safety of Earth's Van Allen belts.) By bringing the tiny craft to 15 to 20 percent the speed of light, it could reach the Alpha Centauri system in twenty to thirty years.

Following their long, hopefully uneventful journeys, such probes could be used to survey and explore uncharted planets and star systems at ever greater distances. Although any effort to report back to Earth would take decades, centuries, and eventually much longer, over the millennia we would acquire and build a vast knowledge of our galaxy. This would allow us to answer many of the great questions about life and intelligence in our universe.

Our sun is an unexceptional G-type main sequence star located in the Orion Arm of the Milky Way galaxy, approximately 25,000 light-years out from its center. A little over 100,000 light-years across, the

Milky Way has roughly 100 billion stars (a challenging calculation to make) with a mean distance of about five light-years between stars, so we'll obviously need a lot of probes to explore it. But if, upon reaching their destinations, each payload of nanobots was to build and launch ten new probes, it would take only eleven generations of ships to achieve that number (10^{11}). Even if each new generation launched only three probes, this would take about twenty-three generations. These calculations assume no ships would be lost, and therefore it would make sense to launch several probes at each iteration, much as a dandelion disperses many dozens of seeds because not all of them will reach fertile ground.

Using this method of exploration, and averaging travel at 10 percent of the speed of light, our entire galaxy could be surveyed and colonized in about 600,000 years.[9] While this may sound like a very long time, consider that the Milky Way has been in existence almost from the beginning of the universe, with an age of about 13.6 billion years. By comparison, our galactic survey would be a drop in the cosmological bucket. Relative to the potential longevity of future intelligences, it might not be so long either.

But our galactic survey needn't stop there. Depending on what the probe discovers, it may begin terraforming the planet it has found. This might take centuries or perhaps even millennia. Fortunately, any new advancements that are made back on Earth could be transmitted to the new probes. Since information can travel at light speed, or about ten times that of our conjectured craft, this would allow the probes to benefit from any knowledge gained during their long period of travel.

The probes could have many other uses as well. Once our intelligence is no longer based on biology and is of a nonbiological nature, we may be able to encode the instructions needed to build bodies, brains, and even knowledge and memories using the tools of nanotechnology. While the amount of information required for this may seem daunting, the development of molecular data storage later this century will provide us with incredible densities. Synthetic polymers have already been developed that when scaled up will be capable of storing one zettabyte (10^{21}) in only ten grams of material.[10] Such density would allow for storing every bit of information in the world, including text, video, and all of the big data created by every corporation that has ever existed, all in only ten ounces of storage.

While it will take an immense amount of data to store the state of a person's cellular structure, including the potentiation and weights of all of their neurons, synapses, and supporting cells, there is also an enormous amount of repetition inherent in these biological subsystems. Additionally, all human beings are incredibly alike, in that we all share 99.9 percent of our DNA. These factors would allow for considerable data compression of whatever aspects of these colonists were still biological, especially if we mixed lossless and lossy compression methods, depending on what features were being stored. While DNA is an incredibly compact method of storage, the information stored is used for the development of a new human being through a very lengthy and highly supervised process. Along the way, there are all manner of environmental conditions and life experiences that must shape that person over the course of decades, which is not necessarily conducive to bootstrapping a distant interstellar colony.

However, using the nanotechnological processes described earlier, it should be possible to one day grow a person in only a matter of hours or days from all of this information. The person would have already been born and grown up on Earth and would have chosen to be part of this mission. To be clear, they wouldn't have been stored as biological DNA but as information contained in ultradense molecular storage. From this information, they would be reconstituted as an adult human, having all of the knowledge and memories they had at the moment they were recorded.

An alternate approach would be to beam information for assembling a person at a certain time *after* the ship left. The idea would be to have a station ready to receive the information sometime after landing. While we have already stretched credulity to its limits, this would require so much energy, error correction, and luck that it boggles the mind. Nevertheless, for civilizations far more advanced than our own, this might be a possibility.

Either of these methods would make it possible to transport entire communities of colonists elsewhere in the galaxy. They would awaken with little sense of the centuries or millennia that had passed to start a new life and a new world. Just as with the earlier examples of cloned minds, this would not be the actual person who acted as their original template back on Earth, even though they would have all of that person's personality, thoughts, and memories. From this moment forward, they would be a separate and unique individual, leading their life in an entirely different part of the galaxy. They would live and hope and dream just as any of us do. Depending on where their new home was located, perhaps on a clear night they might peer up into the dark sky and search for the star that their recently reestablished memory assures them had once been their home.

But what of the other species and intelligences we might encounter? What about *Star Trek*'s "prime directive" that forbids interfering with other civilizations? There is something to this idea; after all, we wouldn't appreciate being on the receiving end of such "colonialism." On the other hand, if we can build technological intelligences that are smart enough to perform the tasks we've just discussed, we can also make them smart enough to avoid certain kinds of interactions. Even if we limited our outposts to only those solar systems where sentient or sapient life was absent, there would still be an enormous number of systems and planets to explore.

Then there's the fact that we don't actually know if intelligent life or technological civilizations exist beyond our own planet. Yes, one interpretation of the statistics is that the universe must be teeming with life. But just what are those statistics? To date, we have a data point of exactly one, and that is our problem. If we had even a little more evidence, we would be able to make a much more informed assessment about how friendly the universe is or isn't to life and advanced civilizations.

Many of us want to believe that the numbers that get plugged into the Drake equation (described in chapter 1) and other similar efforts all but confirm the universe's propensity for life. But until we know otherwise, we really need to treat ourselves as the anomaly we very possibly are. While the mediocrity principle[11] would have us believe we are one of many forms of life, intelligence, and civilization in the cosmos, our existence is just as likely the product of uncanny chance, because we are a self-selecting sample of one. If we are very fortunate, we may one day detect electromagnetic signals of a pattern that denotes intelligence, but this effort makes a lot of assumptions about our universe. The knowledge derived from a close-up and extended survey of our galaxy would provide the data and insight we need to make a true assessment about how common life is or isn't, in all its incarnations.

Which brings us back to Fermi's question: "Where is everybody?" While the universe should be teeming with life and hopefully is, it is unlikely we will ever interact with that life unless it is sufficiently developed and has the desire or need to expand outward from its own little corner of the universe. Overall, the odds are not good.

Consider amino acids, those preliminary building blocks of life. We seem to be able to find them everywhere—comets, asteroids, the far reaches of the galaxy. The autocatalytic processes that lead to polymers, proto-cells, and so forth are probably also ubiquitous, though these latter processes certainly need more time and probably a planet on which they can self-organize and proliferate. From there, we proceed up the slope of complexity to some form of unicellular biology, then perhaps multicellular organisms, all the way to some form of sapient life and perhaps beyond.

But each step is going to be unknown orders of magnitude rarer than its immediate predecessor, and that's a problem. First, without empirical evidence, we don't know how much rarer those later steps might be. Then, assuming other sufficiently empowered technological intelligences do exist in the universe, how rare and how distant from us are they? If they are only a mere five hundred light-years away—a cosmic stone's throw—traveling at our current maximum speed of approximately 165,000 miles per hour, it would take about 2.1 million years for them to reach us—if they knew where we were and wanted to meet us. One might call this expectation the height of human ego and self-importance. Why would anything make such a decision, given the economic, physiological, and psychological challenges involved? Long-term exposure to radiation and microgravity[12] are only two of the enormous perils that would have to be overcome.

No, the challenges of traveling through such vast distances of space will certainly require advanced technological solutions. First, it will probably require forms of intelligence that aren't confined to fragile protein-based bodies that would quickly be destroyed by radiation damage and other dangers of space travel. Second, any space-faring intelligence will need to develop methods of travel that reach a significant fraction of the speed of light. Third, if they are to explore the galaxy in any thorough manner, they will need to have the ability to exponentially expand their explorations through some method of self-replication.

Here, we are already talking about civilizations far in advance of our own. Given the nature of exponential progress, it is possible we are rapidly approaching such capabilities, but this is far from a certainty. Nor, as we saw, is our long-term survival as a species guaranteed. We may be facing a very common bottleneck that few, if any, civilizations in the universe survive. Again, all of this increases the rarity of advanced civilizations being found on any planet.

However common or uncommon such civilizations are, though, it doesn't follow that alien intelligences will share our curiosity and penchant for exploration. In fact, we should probably assume they will not. Additionally, while they may have their own cognitive biases and inclinations, they probably won't share many—if any—of our own, given our entirely separate evolutionary histories. Because of this, they may simply come to the conclusion that such an expedition is a bad idea.

Another factor to consider is gravity. Planets need to have sufficient mass in order to retain enough of an atmosphere for life to evolve and perpetuate. By a number of measures, our own planet sits at the inside edge of our solar system's habitable zone. According to one hypothesis, planets between two to three times the size of Earth are actually better suited to life, a factor that has led to their being dubbed "superhabitable." That is, life would be more likely to evolve on them, assuming that planet's star is long-lived and doesn't emit high levels of radiation, and that the planet itself occupies that system's "Goldilocks zone."

But a planet that is double or triple the mass of Earth would also be that much more difficult to escape. Even using our most powerful chemical rockets, we can barely lift ourselves out of our own gravity well. Based on this, what percentage of advanced life-forms elsewhere in the universe would even be able to leave their planet?

Given these considerations, it seems likely that the vast majority of technology-wielding civilizations will venture out from their home world only for the purpose of survival, whether to seek out additional resources, to extend available living area, or as a backup plan for their own species.

Of course, we might just get lucky. On October 19, 2017, a mysterious object was discovered passing through our solar system. Based on its size, shape, velocity, and angle of travel, it quickly became evident this object had arrived here from interstellar space. As the first detected interstellar object to visit our solar system, it has sparked great interest as to its origins and purpose. Some scientists and laypeople speculated about its being an alien spacecraft or probe, not least because of the unusual trajectory it took. Formally designated 1I/2017 UI, it was more poetically dubbed "Oumuamua," which in Hawaiian means "a messenger from afar, arriving first."

Over time, explanations were formulated, and the notion of its extraterrestrial origin is now being much less seriously considered, but the size of Oumuamua and its distance from Earth ensure we will probably never know for certain. Nevertheless, we have to ask ourselves: if this had been a product of alien technology, would we be capable of recognizing it as such?

Conversely, what if Oumuamua actually was an intelligent probe? If it scanned our solar system as it passed through, can we be certain it would recognize our world as a global civilization? Or would it have concluded that "there's no intelligent life here" and moved on?

Every form of life will be a product of its evolution and therefore will probably be biased about what it does or doesn't classify as intelligence. But if we are able to prepare ourselves in order to recognize this fact in some future extraterrestrial encounter, we may discover the evidence we need to establish our place in the universe, at least to some degree.

Even if this never happens, even if we eventually determine that we are the sole sapient species in all the known universe, that will be okay too. To understand how incredibly unique we are amid all of the grand complexity and entropy that the expanding cosmos has given us would be as meaningful as knowing we are part of a vast community of independently evolved intelligences. Rather than see ourselves as insignificant, as so many people do when they contemplate the unfathomable essence of reality, we will know that we are its crowning achievement.

Or rather, its crowning achievement up to that point. Because it would be up to us, the tiny human beings who once stood on an unexceptional blue planet that circled a typical star on the outskirts of just another spiral galaxy among billions and billions of galaxies, to awaken the universe so that it might one day come to be suffused with the complexity we call intelligence.

CHAPTER 18

THE QUEST FOR POWER

"We live in a world bathed in 5,000 times more energy than we consume as a species in the year, in the form of solar energy. It's just not in usable form yet."

—Peter Diamandis, founder, XPRIZE Foundation

The sun is a blazing jewel set upon the blackness of space. What would otherwise be a blinding brightness is subdued by our viewport filter, transparent to only a restricted part of the visible spectrum. Looking on from our ship, we watch as twenty-third-century solar satellites engage in an elegant ballet around our local star. We're here to watch the installation of the first wave of a new class of solar energy generators. Designed to supply 95 percent of Earth's off- and on-planet energy needs, it is hoped these will be sufficient for our next century's needs and hopefully some time beyond.

One after another, as the small craft take their place on an imaginary three-dimensional grid, each begins to extend a wand made up of thin carbon nanotubes away from the sun. As each wand reaches its maximum length of just over a hundred meters, a further superstructure unfurls from it, along with the thinnest of photovoltaic sheets. More than two acres of the self-healing, electricity-generating material spreads out to receive the full light of the sun. Soon, the surrounding region of space is filled with giant parasols as far as the eye can see.

All of these satellites are about to become nodes for a vast solar generator. As each comes online, it links up in low-power communication with its neighbors. Following a simple set of rules, each node can now stay oriented to the sun, using radiation pressure to maintain its position relative to the neighbors around it. Paths are coordinated to ensure sunlight continues to reach the inhabited planets and outposts beyond the swarm.

As the sun's light is captured and converted to electricity, the satellites relay it from one to the next using laser transmission. This dance of energy continues until it reaches the system's central command module, where it will be transmitted to Earth and beyond, where it can be used to power another century of progress.

It is becoming increasingly evident that any future that continues the cosmological trend toward greater complexity and new forms of intelligence will be increasingly technological. Even while we may retain components of biological intelligence, either individually or as a civilization, it will be rapidly surpassed by that proportion of intelligence that is nonbiological. This trend toward ever greater intelligence will result in exponentially greater demands for energy, which will in turn feed further intelligence and complexity growth, which raises the question: where is all of that energy going to come from, and what can its growth tell us about the intelligences that may eventually develop it?

As we've progressed through these later chapters, the technologies described have become increasingly speculative. Such is the nature of exploring the future: the farther out we look, the more uncertain forecasts become. That said, we have reached the stage in our development as a technological species that we have a very good understanding about many of the physical laws and processes in the universe. We even have a pretty decent understanding about what we cannot know.[1] As a result, we can actually make some fairly reasonable extrapolations about the kind of conditions we will eventually be living in, as well as the demands of our increasingly intelligent civilization.

Perhaps one thing we can be most certain about will be our future demand for energy. This shouldn't surprise us. Looking back to Chaisson's observations about energy-rate density, we see an exponential trend of increased energy usage as we move up the complexity scale. From stars to planets to plants to animals, the amount of energy processed ascends rapidly, and over decreasing scales of time.

Our global society is only continuing this trend. Current world energy use has increased by a factor of twenty-five over the past two centuries, but by far most of that growth has come in only the last sixty years.

A substantial portion of this growth can be correlated to our escalating use of computers. When we factor in cloud computing, the internet, and smartphones along with personal and business computing, energy demand for computers is doubling approximately every year.[2] Based on current global production of 10^{21} joules per year, or about 18 terawatt hours, the Semiconductor Industry Association projects computing will use all of the world's energy by 2040. While this is an absurd statement and a self-limiting condition, it highlights just how rapidly our energy demands for computing are growing. Even if we factor for future energy production growth (the US Energy Information Association estimates growth of 28 percent during the next twenty years) and maximum energy-efficiency gains in all computing processes,[3] this would buy us only an additional decade or so, based on these projections.

Will future growth and energy consumption come to a screeching halt? Or will we find ways around these challenges that allow us to continue the thirteen-billion-year trend? (That's not really a serious question, is it?)

For decades, renewable energy sources have been researched and developed, advancing their efficiencies even as their costs have plummeted. Still, total world renewable energy production is at only about 5 percent. That number is about to see a significant shift. According to the global management consulting firm McKinsey, the cost of new solar and wind production will be cheaper than coal and gas almost everywhere in the world by 2030. This will not only reduce energy costs in a continuing downward spiral, it will allow us to increase production to keep up with future anticipated demands.

We can do so because the sun, which drives virtually all forms of renewable energy, will continue to shine for billions of years. Converting sunlight into energy we can use, primarily in the form of electricity, is the most efficient method currently available to us. As photovoltaic cells become still more efficient, our ability to convert the sunlight striking the Earth into usable energy will increase, but there is only so far it can take us. Assuming demand continues to grow, which it certainly will, energy production cannot keep up with demand beyond a certain point, even if we cover every square inch of the planet with photovoltaic solar panels. In addition to this, we will continue to need some high-density energy sources, notably fossil fuels, for years to come. Eventually, though, we should have alternatives to fossil fuels too.

In light of recent concerns about global warming and climate change, our world's growing demand for energy presents a considerable problem. How can we continue to develop as a civilization without destroying our planet's ecosystem? Looking at existing and projected energy trends, it would seem that our best long-term strategy will be to focus on increasing the rate at which our planet radiates waste heat, rather than endlessly endeavoring to reduce global energy consumption. Lowering the levels of atmospheric carbon dioxide and other greenhouse gases will be one means of achieving this. Reducing the amount of sunlight that warms our planet by increasing Earth's albedo—its reflectivity—might be another approach. Maybe we'll block a small percentage of the sun's energy from entering the atmosphere with orbiting shields or by introducing controllable reflective particulates into the upper atmosphere. Such methods are likely to have a greater chance of success than continually reigning in the energy consumption of our evolving global society.

At some stage in the not too distant future, it will become necessary for us to expand our energy horizons quite literally. One of the first likely candidates will probably be space-based solar power, or SBSP. The concept here is to set up arrays of collectors, probably made up of panels of solar cells and reflectors. The collectors could convert sunlight into electricity, much as they do on Earth. However, they would do it far more efficiently in space because the sunlight wouldn't be filtered by the Earth's atmosphere. This electricity would then be used to generate some form of focusable electromagnetic energy, probably microwaves or lasers. The energy beams would be focused on Earth-based receiving antennas, known as *rectennas*, which would convert the signal back into electricity.

While expensive to set up initially and maintain, SBSP has many advantages over Earth-based solar farms. First, our atmosphere reduces the usable sunlight reaching the ground by approximately 50 to 60 percent. In the vacuum of space, this loss is totally eliminated. Second, weather, the seasons, and nighttime all have a huge impact on electricity production. Again, in space, this really isn't a consideration. Finally, three-quarters of our planet is water, which creates many of its own challenges for establishing solar farms, not to mention that a significant portion of our land needs to remain available for the basics of life, including food production. The usable volume surrounding our planet is vast compared with its surface area, and would be capable of meeting our energy needs for many years to come.

Why will we need so much energy and the computing power it would make possible? For exactly the same reasons we've watched our need for it grow throughout this past century. The electrification of our society was a tremendous paradigm shift, one that accelerated technological progress, even as it escalated our need for energy. Not coincidentally, the advent of the computer age that followed increased our energy demand still further. All of the miraculous future technologies that have been discussed in previous chapters will generate even greater demands. So will the development of new forms of technological intelligence.

As immense as our society's appetite for energy may seem today, it's only the beginning. To understand just where we are relative to where we will be, let us return for a moment to the search for extraterrestrial life and intelligence.

A key consideration in thinking about this search is that while our planet has been emanating radio waves for a little over a century and we began dipping our toes into space travel only about half that time ago, advanced alien civilizations could potentially exist that are millions of years old. If so, then their societies might very likely wield control over incredible amounts of power. They might build megastructures in space the size of Earth's orbit or have the power of black holes at their command. From our current vantage point, such capabilities may feel outlandish or even godlike, just as the power our own civilization commands would have seemed to someone who lived only a few centuries ago.

Thinking about such scales can be challenging. Fortunately, we are far from the first to consider the matter. In 1964, Soviet astronomer Nikolai Kardashev hypothesized a scale of three types of civilization based on the amount of energy they could use and control that subsequently became known as the Kardashev scale.[4] The three types were:

Type I—Planetary civilizations capable of storing and using all of the energy available on that planet, including all of the sunlight that falls on it.
Type II—Stellar civilizations capable of harnessing the total energy output of their local star.
Type III—Galactic civilizations able to access the energy of their entire galaxy.

Given the variability in the amounts of energy based on the different sizes of planets, stars, and galaxies, this is hardly a well-defined scale, but it does offer us some fairly intuitive ways of conceptualizing the extreme differences that are possible. If nothing else, it gets us out of the Earth-centric mindset we have lived with for so long.

In case you're thinking we must be a Type I civilization—we are not even close yet. Some estimates have put us at 0.7 on this pseudo-logarithmic scale, which means we are somewhere near the midway point between a Type 0 and Type I civilization.[5] By some forecasts,[6] we are one or two centuries from attaining Type I status, a few thousand years from Type II, and somewhere between 100,000 and a million years from Type III. However, given that our planet's current annual energy budget is somewhere in the range of the amount of sunlight that strikes the Earth's surface every hour, and factoring the rate at which demand has been growing, we could potentially reach a level equivalent to a Type I civilization sometime in the latter half of this century.

Until our civilization launched the Industrial Revolution in the last quarter of the previous millennium, energy use was historically constrained to our needs for daily survival. Fuel derived from the sun's energy kept us warm, gave us food, and that was mostly it. Then, as our world increased in technological sophistication, so too did our energy needs. So it is probably safe to say there is a correlation between the amount of energy a civilization consumes and its level of technological sophistication. Likewise, the knowledge and access to the means to perform work (using the physics definition) will drive even faster advancements in the future. The exponential growth that ensues from this is likely to continue to surprise us, as exponential trends consistently do.

So, assuming we and eventually other intelligences continue to drive and develop new means of energy production and harvesting, what might we expect to see in the decades, centuries, and millennia ahead?

For the time being, we are mostly limited to accessing forms of energy produced by our sun, whether as a primary source, such as direct conversion using solar cells, or as a secondary one, as with plant-based biofuels and fossil fuels (geothermal and nuclear energy being two notable exceptions). Surrounding our world with space-based solar power will take us to a Type I civilization, perhaps by late this century, but in time our escalating demands will exceed this level of energy production as well.

New technologies will most certainly be needed if we are ever to develop into a Type II world. Some you'll no doubt have heard of, especially if you're an avid reader of science fiction. That's because when science fiction worlds need an enormous energy source for an increasingly off-world civilization, they often turn to a *Dyson sphere*.

A Dyson sphere is a hypothetical structure conceived by theoretical physicist and mathematician Freeman Dyson.[7] As traditionally interpreted in fiction, a massive shell is built to encase a star, often extending many millions of miles in diameter, so that all of a star's energy can be captured and put to use. Unfortunately, this is not feasible, nor is it what Dyson originally envisaged. Such a sphere would require a phenomenal amount of material. If such a structure extended one astronomical unit—Earth's average orbit of 93 million miles—it would have around 600 million times the surface area of our planet.[8] On the positive side, this enormous expanse would give our population plenty of room to spread out. But it would also require all of the rocky planets in the solar system to be broken up and repurposed, including our own. If we're optimistic in our calculations, there might be enough material to make the shell wall between one and two meters thick. Issues such as the shell buckling, gravitational instability, asteroid and comet collisions, shielding from solar radiation, and thermal overheating would probably render this version of the idea unfeasible, if not impossible. However, this was not what Dyson actually envisaged with his concept.

The later-named Dyson swarm would be made up of millions of orbiting collectors that maintain positions relative to each other as well as to the sun, harvesting and transmitting the energy they capture, much as we saw in the opening scenario. These would require far less material to build and could be deployed and brought online more incrementally. A slightly different take on this is the Dyson bubble, which would be made up of collectors using solar sails designed to balance the collectors against the sun's gravity. Dubbed *statites*, these generators would not need engines or fuel to adjust their positions as they orbited the sun. On the other hand, at the times when one statite would eclipse another, the interposition would not only reduce the amount of energy the eclipsed statite could capture, but also alter the solar pressure on its sail, affecting its orbit and position relative to all the other statites. Careful planning would be needed to ensure the rules of swarm behavior didn't result in unanticipated interactions that could eventually lead to a cascade of collisions and catastrophic failure. Collected energy could be passed back through the adjacent nodes in a mesh network to a central location, from where it would be beamed back to Earth and elsewhere in the solar system. Using a branching 3-D fractal arrangement, it would be possible to maximize energy harvesting using minimum resources, much as a plant's vascular system or an animal's circulatory system maximizes efficiency.

Since the Kardashev scale is about the amount of energy society controls, we needn't necessarily obtain it all from the sun's output. Some researchers have established the energy levels for Types I, II, and III as being equivalent to 10^{16} watts, 4×10^{26} watts, and 4×10^{36} watts, respectively. A civilization that builds a sufficient number of planetesimal-sized[9] fusion generators to supplement their energy appetite should also be able to eventually achieve Type II status.

All of this assumes that our civilization survives long enough to develop such amazing technologies. As noted earlier, the next one hundred to one thousand years could be critical to whether or not we succeed. But if we do, these technologies will help us assure our continued survival. In controlling so much energy, we could redirect or obliterate a large asteroid that was on a collision course with Earth. We could reshape any planet or moon in order to make it habitable. We could possibly even change the orbit of the planets themselves, if we wanted to. Perhaps we might transform a planet like Venus from its current hellish lead-melting conditions to something much more similar to our own. But the first trick will be for us to survive long enough.

Type III civilizations would be powerful beyond our imaginings. The energy used by such a society would require incredibly futuristic technology, yet it still remains in the realm of the physically possible. In fact, there are several theoretical technologies we can already conceive of that could generate or harness that level of power.

One of these would be the controlled harvesting of energy from quasars. A quasar is typically found at the center of a galaxy where a supermassive black hole is located. As gas is gravitationally drawn toward the center, it forms an accretion disk that encircles and feeds the black hole, similar to the way water circles a drain before disappearing from sight.

The accretion disk is many, many times larger than the black hole itself, which is spinning extremely fast (due to the laws of conservation of angular momentum), sometimes on the order of half the speed of light! Because of this, the accretion disk, which is also spinning, heats up due to immense gravitational stresses and friction. Reaching billions, even trillions of degrees at its center (remember that our own sun's surface is less than 6,000 kelvins, or 10,000°F), the disk becomes so hot it emits energy across the electromagnetic spectrum, from radio waves to X-rays and high-energy gamma rays. In addition, many quasars emit two opposing jets of plasma that travel at relativistic speeds. This is possibly due to the magnetic fields that are generated around the quasars.[10] The plasma jets can accelerate ionized matter to a substantial fraction of the speed of light and project it for millions of light-years, well beyond its galaxy of origin.

While it has become increasingly evident that the presence of a supermassive black hole is very common at the center of galaxies, the exact mechanism of their formation still remains uncertain. For instance, the supermassive black hole at the center of our own Milky Way, Sagittarius A*, is estimated to have a mass 4.1 million times that of our own sun. But black holes don't grow by simply devouring all matter around them. They are, in effect, very messy eaters, flinging away far more material than they actually consume. Because of this, calculations indicate there isn't enough matter in our galaxy to have formed this supermassive black hole. But if that's so, then how did so much matter accumulate and where did it come from? It's been speculated that when two galaxies collide, the massive black holes at their centers can combine to form even larger supermassive black holes. The collision's disturbance of gas and other matter in those galaxies also accelerates the further feeding of the black hole. The massive influx of matter generates the immense energies produced by the black hole's accretion disk, resulting in a quasar. Just as we can surmise that our own Milky Way must have collided with at least one other galaxy over the course of its existence, projections indicate it will next collide with M31, the Andromeda galaxy, 4.5 billion years from now.

But before we ever tried to tackle these behemoths, we would probably want to practice on something a little less intimidating. Microquasars are many orders of magnitude smaller than the quasars found at galactic centers. Originating from a much smaller spinning black hole of perhaps five to eight solar masses, microquasars are typically orbited by a star, a binary companion. Over time, the companion star will be drawn in by the black hole's warping of space, causing it to orbit ever closer to the overwhelmingly dense mass. But instead of entering the black hole all at once, the star's gas is incrementally stripped from the star to form an accretion disk around the black hole. Some of this material will feed the black hole, while the rest will be accelerated away into space. This process is similar to the way interstellar gas is pulled into black holes at the galactic core, and the accretion disk undergoes stresses that heat it to billions of degrees Kelvin. The more matter that is pulled in, the greater black hole's mass becomes and the faster it spins. Eventually, all of the star's gas is stripped away and much of the output from this cosmic generator shuts down. However, this process might take 10 million years, which would allow it to fuel an advanced civilization for a very long time once the civilization devised the mean to harness this energy. By way of example, the heavily studied microquasar

SS 433 produces two opposing relativistic jets of plasma that travel at a quarter of the speed of light (0.26 c) and stretch out for 130 light-years in each direction. The output of this microquasar would be on the order of 10^{33} joules, or the equivalent of our sun's total annual output, every ten minutes.

Drawing on the energy being emitted at the center of the galaxy would have a far greater payoff. It is estimated that the energy output of our Milky Way's Sagittarius A* is between 10^{37} and 10^{39} watts, equivalent to all the stars in a hundred galaxies, or around a trillion times greater than our sun. Of course, this level of output doesn't come without a price. The largest quasars consume enormous amounts of matter, the equivalent of six hundred Earths every minute! Trying to contain and use these jets of energy would be a huge undertaking. Even the output of a microquasar is far beyond our current abilities to control.

Not surprisingly, there could be other ways to exploit the relativistic effects of a black hole. It is well established that light can't escape from a black hole once the light gets too close and passes beyond its event horizon. Additionally, black holes warp space-time due to their immense mass and gravity. The collapse from a star into a black hole, also known as a gravitational singularity, causes the black hole to spin extremely fast because of the law of conservation of angular momentum. This is because the star that forms the black hole is already spinning. As it collapses, it spins faster and faster, much as an ice skater turns faster when she draws her limbs closer to her body. The tremendous amount of mass that a black hole draws in causes it to spin unimaginably fast, with some of these extreme phenomena spinning faster than three-quarters the speed of light. When discussing black holes, we usually focus on the event horizon, that border beyond which not even light can escape. But some have proposed that there may be a transition region just outside the event horizon that still warps space, but from which escape is still possible. This has been dubbed the *ergosphere*, from the Greek word *ergon*, meaning *work*. This is a region of space that is dragged along by the spinning hole, pulled by its immense gravity. Because it lies just outside the event horizon, it is hypothetically possible to enter the ergosphere and get out again, though this would take a huge amount of energy. Fortunately, the black hole and its ergosphere have plenty of energy available. Since the black hole is not only dragging space-time but doing it at relativistic speeds (that is, at a substantial fraction of the speed of light), there are some novel ways we might steal some of that energy.

One way is known as the Penrose process, conceived by and named for mathematical physicist Roger Penrose. In this process, an object or rocket is launched into a black hole's ergosphere in the direction of its rotation. Here it is accelerated by the rapidly spinning region of space-time it has entered. By firing half of its mass into the black hole, the other half of the object or rocket is accelerated out of it, gaining a boost of momentum from the transaction. Since the remaining half of the rocket has negative-mass-energy under these conditions, conservation of momentum is not violated, because the black hole has given up some of its momentum to do the work. In this way, energy can be extracted from the black hole.

Another, more powerful way we might extract energy from a black hole would be to make use of an effect known as superradiant scattering. For this, a spinning black hole would be encased in an enormous shell of inward-facing reflective material. A laser is then fired into the shell, so it passes through the ergosphere at just the right angle. The beam exits the ergosphere, having gained a big boost of energy. The wall of the shell reflects the beam back into the ergosphere, where the beam gains another boost. As this cycle repeats, a positive feedback loop ensues and the energy beam gains more and more energy until it is finally released. Doing this in a controlled way could extract a phenomenal amount of energy. Of course, there is a catch: wait too long and you create what's known as a *black hole bomb* that would theoretically wipe out an entire solar system or worse.

What kinds of computers or intelligences or civilizations would be powered by all of this? It's safe to say that whatever forms of life and mind exist at that time, they will probably be unfathomable to us. Just as *Australopithecus* could make nothing of a Shakespearean sonnet or an Excel spreadsheet, we would be at a complete loss if faced with the workings of such advanced minds.

Yet there remain things we can still extrapolate. We have explored the idea of advances in technology leading to everything in our environment becoming capable of computation. We've also discussed internalizing our interfaces and using neural prosthetics that would augment our existing biological abilities. We have no reason to expect that the millennia-old trends of enhancing our intelligence with technology will cease.

Technologies that allow us to abstract more and more of the world's complex functions would be a natural application of all of this cumulative energy and processing power. Over the past number of decades, the exponential growth of computers as well as the decrease in the cost of processing power, memory, and storage has created such a surplus that we have had enough to spare for ever more sophisticated and natural user interfaces. Continuing down this road, what powers of the mind, what access to resources might we want to make available to ourselves?

Today's Internet of Things promises to connect all of our devices to the internet, from refrigerators and thermostats to cars and supply chains. But what if that's not enough? Perhaps some future civilization will deem it necessary to be able to access and control every object or every molecule on their planet.

That's assuming civilizations are still confined to planets. A society of digital minds may decide they are better off living in space or on asteroids, perhaps so they can be located nearer to their energy sources. Or maybe they will want to be near whatever communication "trunk lines" exist in that era in order to reduce latency, much as high-frequency traders locate themselves near stock exchanges today. We really can't know where their priorities will be, only that certain trends have consistently driven us in a particular direction for a very long time.

Will the day come when some civilizations will want to impart intelligence everywhere by making everything capable of computation, effectively awakening the universe? Could it eventually make sense that all available resources will be used for this purpose? Perhaps, though it also seems that certain physical laws and realities will preclude this. After all, signals can travel no faster than the speed of light. That could potentially contain this goal of making everything addressable and computable, especially if it takes hours, years, centuries for a signal to travel from one location to another.

Of course, at another level, this is what society has always done. It is just not as apparent today as it used to be when we waited weeks or months for a piece of mail, instead of the milliseconds an email takes today.

Then again, what if a sufficiently advanced society discovers a way to get around this speed limit? Not by breaking the laws of the universe, but by better understanding them? Physicists have long conjectured about the idea of warping space, creating wormholes that would allow us to move between two regions of the universe, seemingly defying the speed limit of light. While it seems highly likely that this would require a phenomenal amount of energy to achieve, we could be pleasantly surprised. It also seems that sending information via a wormhole would require much less energy and be less risky than teleporting a living creature (if the latter is even possible).

One strategy in the search for extraterrestrial intelligence is to look for particular passive energy signatures. This mission is known as Dysonian SETI, and among its proponents is the Glimpsing Heat from Alien Technologies (G-HAT) SETI program.[11] While there may be any number of reasons an alien civilization may not choose to try to communicate with us intentionally, G-HAT and others have reasoned that passive signs of an advanced civilization, such as waste heat, may be detectable, which leads to the idea of searching for hotspots in the infrared range of the spectrum. However, this follows only if we assume an advanced society will allow that level of energy to be radiated. Yet any civilization that commands the energy to build Dyson spheres and Matrioshka brains (see below)[12] would probably have the capacity to be highly efficient with respect to waste energy. While the laws of thermodynamics must be obeyed, the idea of recapturing waste infrared for use as another energy source, either by that civilization or by others in that ecosystem, can't be ignored. Combined with their perhaps explicit choice not to light a beacon by which all the rest of the universe can find them, it would seem that an infrared search strategy such as this has a very low likelihood for success. Perhaps not surprisingly, another paper studying the infrared detection of Kardashev III civilizations concluded that these "are either very rare or do not exist in the local Universe."[13]

On the other hand, given our current understanding of the universe, identifying and tracking phenomena that appear to conflict with our understanding of the universe's natural laws could help us detect technological civilizations well beyond our own. Whether an object that appears to be moving faster than light speed or stellar spectra that fall well outside of expected norms, these could be red flags we would want to examine further.

Continuing along the scale of exotic technologies, we come to computronium, or what MIT professor Seth Lloyd has called "the ultimate laptop." You may recall that computronium formed the alien computer and civilization from the first chapter that was home to one trillion virtual minds before it blew up and destroyed an entire solar system. A hypothetical material of immense sophistication, computronium uses every particle, atom, and quark to perform calculations at the fastest speeds physically possible. As conceived, it would be the densest, most computationally intensive point in the universe. It is constrained by the Bekenstein bound, a theoretical physical limit on the amount of information that can be contained in a finite region of space, as well as other universal limits on computation speed, energy consumption, and the thermodynamics of storage. Such a computer would be 10^{30} times faster than the fastest supercomputers today. That's a million trillion trillion times faster.

Computronium is so dense, it is close to becoming a black hole itself. Because of this, some people have speculated about whether it could be possible for a black hole to actually be a superdense supercomputer, whether as a home to some hyper-advanced virtual society or as a tool for beings who know how to unlock its secrets. Currently, it seems this will forever remain a mystery to us, but who knows if we might one day be able to unlock the answers?

Of course, building superdense computers is only one course we might develop in our faraway technological future. We might also decide to go big. Very big. Some theorists have speculated on the possibility of one day building computers the size of planets.[14] Computers this large would have immense capacity, and the amount of computation would be defined by the scale of the processing components, since speed increases as scale is reduced, largely due to reductions in latency, which is another way of saying the length of time it takes a signal to travel from one point to another. Using optical communication traveling at two-thirds the speed of light (due to slowing in a medium), a Jupiter-sized computer would see latencies from its center to its perimeter in excess of an eighth of a second, or 125 milliseconds. This might be barely tolerable in our present-day communications, but it seems unlikely it would be an acceptable lag time for our more sophisticated descendants. While latency severely limits how much this concept known as a *Jupiter brain* could scale, a planet-sized computer has the potential to provide a tremendous amount of processing power overall.

If latency is ignored, or at least reduced as a consideration, there is another novel computing architecture that would allow for a vast upscaling of computing power, memory, and storage. Building on the concept of the Dyson sphere, a Matrioshka brain is designed to maximize the use of all available energy from a star and turn it into computation. Beginning with an inner shell that surrounds the star, all solar energy across the spectrum, from radio waves to X-rays, would be captured and used for processing. Because computation creates heat, enormous heat in this instance, it needs to be radiated in order for the computer to remain cool enough to function. Even if a system can withstand phenomenal temperatures of millions of degrees, heat loss will be an inevitable byproduct. The waste heat would subsequently be captured by a second Dyson shell arranged outside of and at some distance from the first shell and converted to a usable form, presumably electricity. The process can be repeated as far as it makes sense to its builders. Presumably, no matter how advanced the civilization, there would come a point after which the cost-benefit analysis wouldn't make sense, either in terms of finances, energy, or the material resources required to build such enormous structures.

But what if the most advanced processing didn't need any material at all? Is it possible that we or some other intelligence in the universe could one day transform ourselves into beings made up of nothing but light, of pure energy? It's hard to imagine, since it has long been assumed that light doesn't work that way. Since photons are massless, it has typically been thought that there is no way for them to interact with each other in any significant manner. But there are hints of mechanisms that might eventually make this possible.

In the past few years, a new form of energy-matter called *photonic molecules* has begun to be explored. This involves photons that are bound together in a strong interaction that aligns their properties so that the photons are virtually indistinguishable from each other.[15] Though made up of two or more ordinarily massless elementary particles, the resulting bound "molecule" behaves as though it actually has mass. As a result, bound photons move 100,000 times slower than the speed of light. This state of being midway between matter and energy has led to comparisons to the light harnessed by lightsabers in *Star Wars*.

A related phenomenon that isn't quite "all energy" is known as photon pair production. When high-energy photons pass near an atomic nucleus, a pair of subatomic particles—specifically a particle and its antiparticle—are produced. The pair might be an electron and a positron, or a muon and an antimuon, or a proton and an antiproton. This creation of matter is possible because energy and mass are equivalent. The ability for energy to create matter and matter to become energy is precisely why $E = MC^2$ (Energy equals Mass times C— the velocity of light—squared). There is a direct relationship between the two, which is why a nuclear chain reaction in matter can release such tremendous amounts of energy. Conversely, it takes a very high-energy interaction to create new matter.

While much of this research is discussed in terms of its hopeful and eventual application to quantum computing and communication, other possible future technologies come to mind as well. As the ability to manipulate energy with energy develops, could it eventually become possible to create self-sustaining "energy structures"? Created entirely from photons, or using transitory forms of matter produced by energy, these could allow for computation to be performed, and from computation comes the potential for intelligence and eventually mind. But even the basic experiments of today require extremely high energy levels to create the needed conditions, so imagine how much energy might be required to form an "energy mind"!

The minds and civilizations of the future will be unlike anything we have experience with today. As a result, it is highly challenging to anticipate what forms they may take or what values they might have centuries and millennia and eons from now. But we can look to the trends of the past and present in order to gauge certain aspects about the future, not least regarding energy.

Energy has driven our universe from the very outset. Before there was matter or chemistry or information or intelligence, energy was there. And from those earliest beginnings, the universe of energy has been running down, becoming increasingly more disordered, more chaotic. Relentlessly. Moving in a single direction that will one day result in the end of everything everywhere. As we are about to discover.

CHAPTER 19

LIFE AT THE END OF THE UNIVERSE

"Just as the constant increase of entropy is the basic law of the universe, so it is the basic law of life to be ever more highly structured and to struggle against entropy."
—Václav Havel, Czech statesman and writer

Our ship slows as it approaches the outer reaches of the ultra-massive black hole. All around us, planets, ring worlds, Dyson spheres, space stations, and every manner of smaller craft occupy this region of space, spreading out in a volume that would have dwarfed old-Earth's solar system.

We've traveled 100 trillion years into the future, and all of the stars in the universe have long since disappeared from view and died out. All that is left in the universe are an unknowable number of ultra-massive black holes,[1] the last sources of energy in all the cosmos. Our destination is Beyonfar, an extended spaceport orbiting the only black hole within our light cone, what is for us the observable and accessible universe.

We assume a position near an immense spindle-shaped space station, where tractor beams draw us in to an awaiting dock. Departing our ship, I'm immediately greeted by a tall androgynous humanoid, its feet hovering several inches off the ground.

"Fred!" I shout, happy to see my friend. "It's great to see you."

Fred replies in a dulcet baritone. "Likewise, Richard. What brings you to the end of the universe?"

I explain the reason for this recent journey, starting from our exciting visit to the Big Bang. But Fred has heard it all before. Quite literally. Over the course of several trillion years, Fred has probably greeted several billion visitors.

Let me add that obviously "Fred" isn't my host's name. His collective species doesn't have one. So I just choose to call them "Fred.[2]" They have been the de facto hosts at this final waypoint for trillions of years, using the most minuscule part of their attention to welcome, host, and entertain guests from all over space and time.[3] Part tourist destination, part oasis, and last stronghold against entropy, Beyonfar is, for this and many other reasons, the place to be.

Nor is Fred really an androgynous humanoid, or any other form of material being, for that matter. They are an energy intelligence far beyond the complexity of our own species, and we are able to interact and have this conversation only because they deem it worthwhile to abstract themselves many times over. To put it bluntly, they are dumbing themselves way down not only so I can understand them but so I can see and interact with them at all.

Fred's energy species does all of this not just for altruistic reasons but out of curiosity as well. Because of their command of physics around this black hole, intelligent species from across the known worlds have resettled here as a means of survival. Many of the species are our direct descendants. Others may be of entirely alien lineages. Or possibly not. It depends on who you ask.

As for the curiosity part, I think Fred sees us a little like living fossils or museum pieces. Beyonfar gives them nearly endless opportunities to observe the many intelligences the universe has evolved. Are they researchers or collectors? Perhaps a little of both, but I suspect their actual motives are so far beyond our comprehension that any further elaboration would be meaningless.

As Fred and I reach my destination, my host nods and hovers away as I open the door to Fermi's Taproom. Inside, the establishment is bustling with activity, filled with every form of alien life and intelligence imaginable.[4] Spotting my friends, a couple of tall, graceful Tau Cetians at the counter, I go over and soon we're chatting like we'd only just seen each other yesterday. As you might have guessed, this isn't my first visit to the bar at the end of the universe.

Turning to the owner-bartender, a shimmering Kentauran of indeterminate age, I call out over the chatter, "Great to see you, Fermi. I'll have the usual!"

––––––––––

It almost feels like we've come full circle, doesn't it? We started off with a superdense singularity and nothing else, and that's pretty much where we find ourselves here at the end of our story, as well. But let's be clear: what began with a bang could well end in the longest of drawn-out whimpers.

From the very beginning, the universe has been expanding, cooling, becoming less dense and less organized with every passing second. This process has never stopped, nor will it ever. Assuming the laws of physics and our understanding of the mass of the universe are correct, the universe will keep expanding forever. True, there are other theories that say this expansion will reverse itself, resulting in everything collapsing in a Big Crunch. Others think that this might be part of an oscillating cycle and the universe always has and always will bounce from one to the other, an idea that has come to be known as the Big Bounce. And there are other theories. But for my money, indications are we're heading for a very long, very slow heat death that will one day see nothing left in the cosmos, not even subatomic particles like protons.

However, *slow* is definitely the word. Here at the fictional-maybe one day location of Beyonfar, we've been visiting one of the last concentrations of matter anywhere for as far as we can see, and that's a very long way. Space has been expanding for so long and so rapidly it has literally taken every galaxy farther away than its light has had time to reach us. Think what that means. Here, 100 trillion years after the Big Bang, every galaxy is farther than 100 trillion light-years away from us! Actually, it is a lot more than that because as that light has been traveling, space itself has been expanding that much farther.

Stepping back a bit to only three trillion years after you read this, the only stars that will be visible from Earth will be those in our own galaxy. Except that Earth won't be around by then. Our planet will have long been swallowed up by our sun, which will have expanded into a red giant, some seven and a half billion years from now. Our own Milky Way galaxy will still be relatively gravitationally bound at that point, but that doesn't mean it won't see hard times. Again and again, it will crash into M31, the Andromeda galaxy, colliding with and altering the relative positions of the billions of stars in each. But by the time of Beyonfar, 100 trillion years in the future, all of the stars in all of the galaxies will have long burned out. More importantly, there will be insufficient material anywhere in the universe to naturally form any new ones.

The ultra-massive black hole from our scenario will exist for far, far longer than any of these other celestial objects. In the early twenty-first century, the largest known ultra-massive black hole is at the center of the quasar TON 618, with a mass equivalent to 66 billion times that of our sun. This results in an event horizon—the point beyond which light is unable to escape—with a radius of 1,300 AU. (Recall that one AU is nearly 93 million miles, the radius of the Earth's orbit, so this black hole's event horizon, also known as its Schwarzschild radius, is more than a hundred billion miles from its center.) Presumably, the black holes of this scenario will be far more massive than even this monster, so you can only imagine how far their event horizons might extend.

But large or small, black holes won't last forever either. Over time, the phenomenon theorized by physicist Stephen Hawking, called

Hawking radiation, will cause these black holes to essentially evapo-rate, incrementally losing mass until they cease to exist. But this will be an extremely lengthy process. It is estimated that the last black hole will have evaporated away 10^{100} years from now. (Our fictional Beyonfar is only 10^{14} years after our current era.)

So, while the stars may be long gone and most of our home worlds will be ash and dust, it will still be possible for new emergences to appear and new intelligences to come into being for many, many times longer than the universe will have already existed.

Emergence in this far-flung future would happen very differently than it does today. As we saw earlier, shine enough sunlight or gen-erate enough heat from an undersea thermal vent, and the simplest of molecules start forming into amino acids and autocatalyzing poly-mers. A mere few hundred million years later, life may appear, and in the flicker of a few billion years more, you could have a spacefaring technological society.

But in an era of civilizations that are able to command the power of black holes as nature's last remaining energy source, things would have to be very different. We can't, of course, anticipate the behaviors and motivations of future intelligences, but just as we have a pen-chant for studying our own origins, so some of tomorrow's minds may want to as well. Using the vast energies at their command, cre-ating or simulating emergent behaviors may offer insights into our shared nature. Better yet, perhaps a science of applied emergence will develop, allowing for the exploration and directed development of new emergent behaviors and intelligences. Though this could easily be suited to earlier-stage minds, including our twenty-first-century selves, it seems safe to say that some future species may one day have far more time and energy on their hands than we do!

But why stop there? As we saw in the scenario at the beginning of chapter 14, a sufficiently advanced species may choose to simulate new worlds on a computer. This would allow them to explore all sorts of macro and micro phenomena, either in real time or at vastly altered speeds.

Alternately, it is also possible they could build or populate real worlds in order to observe different species' behaviors and responses. Even as far in the future as our fictional Beyonfar, it should be possible to find the needed material for building new planets from the husks of those burnt-out stars that have managed to evade the fate of being devoured by a black hole. Fragment by fragment or atom by atom, these stars could be disassembled and relocated for the purpose of building new Earths. Not surprisingly, it has been speculated that our own twenty-first-century Earth could be just such an experiment, though given the lack of evidence and how early we are in the overall lifetime of the universe, this seems highly unlikely.

Perhaps the most disturbing of these conjectures is the idea that we are all living in a universe-scale computer simulation. Popularized by the *Matrix* movies, the idea has gained ground among a number of theorists and respected scientists.[5] If true, this would undermine an enormous number of our ideas and philosophies about what we call reality. While it may seem outlandish, recall that virtual environments are already being used to "raise" and test developing AIs in our present era.[6] Isn't it at least possible that we could be part of a lab experiment performed by an extremely advanced civilization of researchers?

To show how seriously this notion of a simulated universe is being taken, academic papers have been published and experiments run to try to prove or disprove it. For instance, in a 2012 paper, a team of Cornell researchers devised a way to use cosmic rays to reveal the underlying structure or lattice of the universe, if one exists,[7] the idea being that if we are in a simulation, our reality must have an underlying matrix or structure on which it is constructed. According to mathematics and the physical laws of our universe, there is a minimum possible size for this lattice, and cosmic rays could be used to reveal whether one exists.

Then, in 2017, a team of theoretical physicists from Oxford University in the UK published a paper[8] that showed that certain quantum effects are actually impossible to simulate in a classical computer. This is because as the number of particles being simulated increases, the complexity of calculation grows exponentially, rapidly becoming impossible to compute. Some people felt this conclusively disproved the idea of our living in a simulation.

Of course, that is not the end of the story. Besides the fact that the paper's authors were not actually working on what is known as the simulation hypothesis, their work focused entirely on using classical computers. No consideration was given to the idea of using quantum computers, which in theory could one day be able to keep up with the exponential growth in such a simulated system's complexity. An even more challenging criticism is that there is nothing that requires the laws of *our* universe to be the same as those of the universe performing the simulation. It seems quite possible that this enigma will never be proven or disproven, unless we are one day permitted to look behind the curtain. Until then, it is likely to remain in the realm of belief systems, with no logic able to reveal the truth.

As future species endeavor to create new, never-before-seen intelligences and environments for them to live in, there is another approach that could yield very "familiar" results: creating a new universe.

Since the 1970s, physicists have explored ideas about what it would take to generate a new universe, as part of their quest to understand our cosmos and the Big Bang. One of the most theoretically promising ideas centers around a particle, predicted by the Standard Model of particle physics, known as a magnetic monopole. The particle was first proposed by physicist Paul Dirac in 1931 but has remained totally elusive, despite several formal searches.

For context, a standard magnet—in fact, every magnet we have ever made or found in the universe—is a dipole magnet, meaning it has a north and south magnetic pole. Neither exists without the other. Cut a magnet in two and each piece now has its own north and south pole. But a monopole is very different. An ultraheavy particle, it is perhaps best described as a spherical knot in the fabric of space.

Several ideas have been suggested as to why monopoles haven't yet been detected, including the possibility that they don't exist. But if we ever do find and capture one, we may have an essential ingredient for generating a brand-new universe. Even if we are not able to locate a monopole, it is possible we could make one in a particle accelerator. Alternately, it has been theorized that some other forms of exotic matter could be created that would have similar properties.

Once obtained, bombarding a monopole with high-energy sub-atomic particles would in theory cause it to generate a tiny wormhole. But, since we would only see the wormhole's mouth, it would appear to us as a miniature black hole. At the other end of that wormhole, however, a new universe would have begun to form and expand, possibly much as ours did at the time of our own Big Bang.

But lest you worry this new universe might expand into our own universe and destroy it, there's a twist. According to theory, the expansion of the interior of that monopole would be its own space, entirely separate from and inaccessible to us in our universe,[9] the exception being that the tiny wormhole connected to it would be a temporary tunnel between our universe and the newly formed one. Unfortunately (or perhaps fortunately), this wormhole would close up and disappear shortly after the new universe formed.

While it's true that we haven't actually found a monopole and we can't manufacture baby universes yet, there is an awful lot of future for us to explore this in, if we can just survive as a species for a few more centuries. High-energy particle physics will become much more powerful in the millennia ahead, as will our knowledge and theories about the laws of the universe.

As if all of this weren't mind-bending enough, it doesn't stop there. According to some grand unified theories,[10] the universe should be teeming with monopoles. There are several possible explanations why this might not be the case, including the idea that our universe passed through its inflationary epoch before temperatures had cooled enough for monopoles to form. In which case they may exist, just in far fewer numbers than has previously been calculated. But if there are all of these potential baby universes floating about inside our own universe, and all of these have their own populations of monopoles,[11] then our universe could be among multitudes upon multitudes of universes that exist, all entirely separate from each other and beyond our ability to ever communicate or interact with.

Such multiverse concepts come up in other areas of physics, such as string theory and M-theory. The multiverse is an amazing concept, but because by definition none of these could be interacted with or observed from our universe, they remain empirically unfalsifiable, a critical standard in scientific terms—there is no experiment that can prove they do not exist, and so they dwell in the realm of belief. (At least for now.) Outside of our imaginations, we will probably never know much about any universe other than our own.

Many people question whether we even have the right to create a new universe, should we ever acquire the ability to do so. Of course, there are the theistic concerns based on the idea we would be playing God, but such arguments go back to Galileo and long before. Then there are those who question whether we have the right to set off a chain of events that will one day result in intelligences and life that could experience suffering. On reflection, though, this is analogous to deciding to bring a child in the world. Yes, they will know pain and pleasure, joy and sadness. These conditions seem inextricably linked to the nature of existence.

Future of Humanity Institute researcher Anders Sandberg suggests there may actually be a moral obligation to generate new life, new intelligence, and new universes. Part of Sandberg's reasoning is that life strives to make its environment better for itself, resulting in a trend that tends to make the universe an overall better place.

I would add that without life and intelligence, the universe is meaningless. This is not meant to be a nihilistic statement, but a matter of subjectivity, because without subjective observation between different aspects of the universe, meaning cannot exist.

In order to give others the near infinite possibilities and opportunities that have been afforded to all of us in our reality, future intelligence may see it as a moral imperative to generate new universes. The only reason they could do this, and that we can live and you can read these words, is because our universe exists. Who are we to deny the same to unknown trillions of other potential universes, worlds, and civilizations?

If we assume that Fred, and perhaps others around Beyonfar, have come to similar conclusions, then they are very likely using some of that tremendous power they are siphoning off from the ultra-massive black hole to generate new universes. Do they suffer some form of compunction from the act? Perhaps it doesn't result in a moral dilemma for them whatsoever. Or maybe for them this act of *cosmogenesis* is a truly spiritual undertaking, a selfless deed that leads to so much, yet will forever remain isolated from and unknowable to them. The ultimate version of "do unto others as you would have them do unto you."

In many respects, this is what nature has been building up to all along. As all of these radical emergences have increased future freedom of action, they have accelerated the cosmos's progression toward equilibrium. At the point intelligence is no longer able to further this for itself in this universe, at least it may be able to perpetuate the process in alternate universes—a final stage of emergences that carry entropy's legacy on into yet another incomprehensibly complex web of cycles.

Let's revisit the nature of entropy and equilibrium, as it relates to Chaisson's interpretation of energy rate density, one last time: if each new level of emergences that results from nature's growing complexity processes energy at ever higher rates, then in a very real sense we may eventually be hastening the demise of our universe. To be clear, if future emergences continue to only passively access the energy produced by cosmic phenomena, that is not going to increase the universe's overall rate of decline. But if advanced future civilizations somehow develop the means of speeding up those energy production rates—whether of stars, quasars, or black holes—in an effort to feed their unending hunger for energy, it may do just that. How much of an effect accelerating entropy production ultimately might have on the equilibrating of the universe, only time will tell.

While the notion of hastening such tremendously powerful processes may seem incredibly unnatural to us from our current vantage point, this needn't always be the case. In conversation, Chaisson observed, "If an advanced civilization were someday able to boost a star's luminosity, wouldn't that civilization still be part of Nature? Since we are a part of Nature, and not apart from it, even some futuristic and fantastic use of energy would have to be considered, I think, not beyond but rather in accord with the Universe's natural rate of energy usage."

Obviously, we don't know nature's "natural rate" of energy production, and based on the evolving character of the cosmos, that rate probably isn't fixed either. So, if an advanced civilization accelerates those processes in order to feed its energy needs, isn't that just a continuation of this long, ongoing pattern?

To our tiny species on our tiny planet, just learning to escape its gravity well, the idea of our impacting the balance of energy in the universe, much less accelerating the cosmos's end, feels like it should be impossible—perhaps as much as our being able to impact Earth's global climate and ecology must have seemed to our ancestors only a century or so ago. But if our descendants do eventually gain the ability to wield and process energy on increasingly larger, ever accelerating scales in the millions, even billions of years ahead, then it might actually be possible.

All of this may feel disheartening, but it's not. It is little more than business as usual, and for all intents and purposes it is the natural order of the universe. We humans are here because countless unlikely events occurred in the past that allowed us to evolve from less complex, less energy-hungry processes. The emergences that descend from us and our civilization in the future will do the same. It is the cycle of cosmic evolution that has taken place from the moment of the Big Bang. From all appearances, it will continue for as long as there is energy to harvest somewhere, in order's continuing struggle against entropy.

To say the universe is vast is perhaps the greatest of all understatements. By comparison, our planet and the people on it are infinitesimal almost beyond measure. We are far closer in size to subatomic particles[12] than we are to the scale of the cosmos. Small wonder then that we often feel so insignificant when contemplating the wonder and grandeur of it all as we stare up into the night sky.

But we are anything but insignificant, because we represent a level of complexity and intelligence that sets us apart from everything else in the known universe. Whether we are alone in all of this or there are many other forms of intelligence out there, alien to us, doesn't ultimately matter. We are decidedly members of the most exclusive club the cosmos has to offer.

Though we may feel at times that all of this vastness, all of these impersonal forces strip any and all meaning from life, think again. The universe has meaning and purpose to be sure, but it is not inherent. For all its power, a quasar cannot reflect on its true nature. Though it may be bathed in sunlight, an autocatalytic molecule cannot appreciate a sunset. While a single-celled organism is surrounded by kindred cells, it is incapable of feeling love for any one of them. Yet through the course of our daily lives, we rarely seem to recognize how extraordinary these experiences are relative to *everything* else in nature.

This is what makes us truly, uniquely special. Because *we* are what gives the universe meaning. We are the witnesses to its becoming. While there may be other witnesses elsewhere in time and space, for now, as best we know, this responsibility, this *duty* is solely ours. This is our purpose and will be our legacy. To fill the cosmos with intelligence, infusing it with meaning that it would otherwise have been denied.

What will *Homo superior*, or *Homo technologicus*, or whatever we eventually become, think of all of this? Will they come to grasp the purpose of the cosmos in ways we never could? And how will they look back and see us? As a primitive curiosity? As a revered ancestor? Or possibly as something in between?

Few things about the future are certain, but as far as we know, the physical laws of nature are immutable. The cosmos will continue running down as its entropy increases. Probability will keep driving complexity, which in turn will lead to new and unforeseeable emergences. Some of these emergences will become organizations of matter and energy that will maximize their future freedom of action, so they may better extend themselves and the emergences that may come after them. All of it in order to perpetuate this amazing, ongoing process that is our intelligent future.

ACKNOWLEDGMENTS

Just as the world we know would not exist without aggregating many different forms of intelligence, this book would not be possible without the generosity, guidance, and assistance of many very talented minds. Though writing is often spoken of as a solitary process, producing a book like this is truly the work of many, and so I wish to thank everyone I'm able to and hope anyone I neglect to acknowledge by name will understand.

The research for *Future Minds* took place in several phases spread out over a number of years. As far as I'm concerned, one group that rarely gets enough credit are the many scientists and academics whose work provides insight and enlightenment for all of us. So much scientific research takes much longer than is realized but is essential to our civilization's incremental building of knowledge. The names and work of many of these researchers are documented in the Notes section, and I'm grateful for their enormous collective effort.

Many scientists and theorists were very generous with their time, speaking and corresponding with me, answering my questions, and helping me better understand their fields and findings. From the field of artificial intelligence, I want to express my gratitude to Yoshua Bengio of Université de Montréal, Cynthia Breazeal of MIT Media

Lab, Joanna Bryson of University of Bath, Oren Etzioni of the Allen Institute for Artificial Intelligence, David Gunning of DARPA, and Gary Marcus of New York University for their invaluable assistance. I'm indebted for the cosmological and complex systems thinking perspectives of Eric Chaisson of Harvard-Smithsonian Center for Astrophysics, David Christian of The Big History Project, David Krakauer of the Santa Fe Institute, and Alex Wissner-Gross of Harvard University. From the field of affective computing, I am deeply grateful for the insights and support of Rana el Kaliouby and Gabi Zijderveld of Affectiva and Rosalind Picard of MIT Media Lab.

Additionally, my thanks to Ben Goertzel of Novamente, José Hernández-Orallo of Universitat Politecnica de Valencia, Paul Nelson and Joanna Masel of the University of Arizona, Steven Pinker of Harvard University, and Vernor Vinge, the computer scientist and science fiction author who popularized the concept of the technological singularity. I appreciated the opportunities to chat about chatbots made possible by Kate Bland and Nate Michel of Amazon and Zach Johnson of Xandra. Thank you also to Qing-yu Cai of the Wuhan Institute of Physics and Mathematics in China for discussing their mathematical proof of how the universe may have spontaneously generated from nothing, a mind-bending idea I think most of us will agree.

I'm tremendously appreciative of all the people who made it possible to actually publish *Future Minds*, beginning with my agent Don Fehr and the team at Trident Media Group. Don connected me with Skyhorse Publishing, where everyone has worked so hard to bring this book to fruition. I'm especially grateful to my editor, Cal Barksdale, whose meticulous work and attention brought clarity and polish to this book. I consider myself a better writer from our collaboration and will strive to build from what he's taught me. Cover designer Erin Seaward-Hiatt and copy editor Katherine Kiger have also helped make my work shine, for which I am thankful.

A big thank you to my advance readers, including Cindy Frewen, Glen Hiemstra, Alexandra Levit, and Gideon Rosenblatt. Finally, none of this would have been possible without the support and understanding of my family, especially my incredible wife, Alex, who regularly reminds me that there are so many ways to view, experience and appreciate this beautiful world we all live in.

NOTES

Preface

1. Marvin Minsky used the term "suitcase word" when talking about terms that have many different meanings that tend to become conflated. Words like "intelligence," "consciousness," "memory," "intuition," and so forth.

2. This quote is difficult to definitively confirm, though Einstein did say something similar in a *Saturday Evening Post* interview in 1929: "Knowledge is limited. Imagination encircles the world."

Chapter 1

1. *Star Trek TOS: The Original Series*, which ran from September 6, 1966, to June 3, 1969.

2. According to interviews recapped in "Where is Everybody—An Account of Fermi's Question" by Eric M. Jones, Konopinski, Teller, and York each recall slightly different versions of Fermi's famous observation.

   Eric. M. Jones, "'Where Is Everybody?' An Account of Fermi's Question," Los Alamos National Laboratory, March 1, 1985.

3. Drake has stated that he didn't intend to calculate the number of alien civilizations, so much as stimulate conversation about their extent and possibility.

4. The mediocrity principle is the idea that any randomly selected item is more likely to belong to a more numerous category than to a category that is less so. In other words, it basically says we aren't that special.

According to this principle, we should find lots of other intelligent life in the universe.

5. Thomas Nagel, "What Is It Like to Be a Bat?" *The Philosophical Review* 83:4 (1974), 435–450, doi:10.2307/2183914.

6. Shane Legg and Marcus Hutter, "A Collection of Definitions of Intelligence," *ArXiv.org*, June 25, 2007, arxiv.org/abs/0706.3639.

7. Shane Legg and Marcus Hutter, "Universal Intelligence: A Definition of Machine Intelligence," *Minds and Machines* 17:4, (2007), 391–444. doi:10.1007/s11023-007-9079-x.

Richard Yonck, "Toward a Standard Metric of Machine Intelligence," *World Futures Review* 4, no. 2 (2012), 61–70. doi:10.1177/19467567 1200400210.

8. A. D. Wissner-Gross and C. E. Freer, "Causal Entropic Forces," *Physical Review Letters* 110, no. 16 (2013), doi:10.1103/physrevlett.110.168702.

Chapter 2

1. Lawrence M. Krauss, *A Universe from Nothing: Why There Is Something Rather Than Nothing* (New York: Free Press, 2012), 183.

2. Dongshan He, Dongfeng Gao, Qing-yu Cai, "Spontaneous Creation of the Universe from Nothing," *Physical Review* D 89, no. 8 (2014), doi:10.1103/physrevd.89.083510.

3. Richard Yonck, "Is All the Universe From Nothing?" *Scientific American Blog Network*, May 22, 2014, blogs.scientificamerican.com/guest-blog /is-all-the-universe-from-nothing.

4. During the inflationary epoch, the universe is thought to have expanded by a factor of 10^{26} linearly. As a volume, this would result in an increase by a factor of 10^{78}.

5. Redshift is the relativistic "stretching" of photons (electromagnetic energy) as they race across the still expanding cosmos. Since it is impossible for light to travel faster than approximately 300,000 kilometers per second in a vacuum, the expanding universe "stretches" the signal, resulting in longer wavelengths, which in turn correspond to a lower frequency. This effect occurs throughout the cosmos, and allows us to gauge how distant different parts of the universe are from us today.

6. It is surmised that dark matter played a crucial role in drawing matter together during this time. However, there is still so much uncertainty about this and dark matter in general that there is limited value in conjecturing on it here.

Chapter 3

1. R. Landauer, "Irreversibility and Heat Generation in the Computing Process." *IBM Journal of Research and Development 5*, no. 3 (1961), 183–91. https://doi.org/10.1147/rd.53.0183.
2. E. J. Chaisson, *Cosmic Evolution: The Rise of Complexity in Nature* (Cambridge: Harvard University Press, 2002).

    E. J. Chaisson, "Energy Rate Density as a Complexity Metric and Evolutionary Driver," *Complexity 16* (3) (2010), 27–40. https://doi.org /10.1002/cplx.20323.
3. Chaisson, *Cosmic Evolution*.
4. Jeremy L. England, "Statistical Physics of Self-Replication," *The Journal of Chemical Physics* 139, no. 12, 2013, 121923. doi:10.1063/1.4818538.
5. Axel Kleidon, "Life, Hierarchy, and the Thermodynamic Machinery of Planet Earth," *Physics of Life Reviews* 7, no. 4, 2010, 424–460, doi:10.1016/j.plrev.2010.10.002.
6. Harold J. Morowitz, *Energy Flow in Biology: Biological Organization as a Problem in Thermal Physics* (University of Michigan: Academic Press, 1968).
7. Cosmic ray spallation and supernova nucleosynthesis are mostly responsible for the elements heavier than iron.
8. José I. Cardesa, Alberto Vela–Martín, and Javier Jiménez, "The turbulent cascade in five dimensions," *Science 25*, August 2017: 357, issue 6353, 782–84, doi: 10.1126/science.aan7933.

Chapter 4

1. A twenty-second genetically encoded amino acid, pyrrolysine, was discovered in 2002. It is used in the biosynthesis of proteins in some bacteria and archaeans, but not in eukaryotes.
2. John R. Cronin and Sandra Pizzarello, "Amino Acids of the Murchison Meteorite. III. Seven Carbon Acyclic Primary α-Amino Alkanoic Acids1," *Geochimica Et Cosmochimica Acta* 50, no. 11, 1986, 2,419–2,427, doi:10.1016/0016–7037(86)90024–4.

    Philippe Schmitt-Kopplin, et al. "High Molecular Diversity of Extraterrestrial Organic Matter in Murchison Meteorite Revealed 40 Years after Its Fall." *Proceedings of the National Academy of Sciences* 107, no. 7, 2010, 2,763–2,768, https://doi.org/10.1073/pnas.0912157107.
3. Bill Steigerwald, "NASA Researchers Make First Discovery of Life's Building Block in Comet," NASA Goddard Space Flight Center,

August 17, 2009. https://www.nasa.gov/mission_pages/stardust/news/stardust_amino_acid.html

4. Kathrin Altwegg, et al. "Prebiotic Chemicals—Amino Acid and Phosphorus—in the Coma of Comet 67P/Churyumov-Gerasimenko," *Science Advances* (May 27, 2016), doi: 10.1126/sciadv.1600285.

5. Arnaud Belloche, et al. "Detection of a Branched Alkyl Molecule in the Interstellar Medium: Iso-propyl Cyanide," *Science* 345, issue 6204, 1,584–1,587, September 26, 2014, doi: 10.1126/science.1256678.

6. Urey was awarded the Nobel Prize in Chemistry in 1934 for his discovery of heavy hydrogen, which is now known as deuterium. He declined to attend the award ceremony in Stockholm in order to be with his wife at the birth of their daughter.

7. Oparin in 1924 and Haldane in 1929.

8. Adam P. Johnson, et al. "The Miller Volcanic Spark Discharge Experiment." *Science* 322, no. 5900, 2008, 404, doi:10.1126/science.1161527.

9. Stuart A. Kauffman, "Autocatalytic Sets of Proteins," *Journal of Theoretical Biology* 119, 1–24, 1986, https://doi.org/10.1016/S0022-5193(86)80047–9.

10. Wim Hordijk, Mike Steel, Stuart Kauffman, "The Structure of Autocatalytic Sets: Evolvability, Enablement, and Emergence," *Acta Biotheoretica 60*, 379–392, doi:10.1007/s10441-012-9165-1.

11. A ribozyme is an RNA molecule capable of catalyzing biochemical reactions, much as an enzyme does for metabolic processes.

12. An amphipathic molecule has both polar and nonpolar parts which determine how it will interact with other molecules.

13. J. M. Berg, J. L. Tymoczko, and L. Stryer, "Phospholipids and Glycolipids Readily Form Bimolecular Sheets in Aqueous Media," in *Biochemistry*, 5th edition (New York: W. H. Freeman; 2002), section 12.4, https://www.ncbi.nlm.nih.gov/books/NBK22406.

A lipid bilayer consists of an aggregation of phospholipids arranged as long inward-directed insoluble hydrocarbon chains joined to outward directed phosphate head groups that are water soluble.

14. Andrew M. Turner, "An Interstellar Synthesis of Phosphorus Oxoacids," *Nature Communications*, vol. 9, no. 1, 2018, doi:10.1038/s41467-018-06415-7.

15. Carl R. Woese, George E. Fox, "Phylogenetic Structure of the Prokaryotic Domain: The Primary Kingdoms." *Proceedings of the National Academy of Sciences of the United States of America*. 74 (11): 5088–5090 (1977). https://doi.org/10.1073/pnas.74.11.5088.

C. R. Woese, et al., "Towards a Natural System of Organisms: Proposal for the Domains Archaea, Bacteria, and Eucarya," *Proceedings of the National Academy of Sciences*, US National Library of Medicine, June 1990, www.ncbi.nlm.nih.gov/pubmed/2112744.

The idea of a third "domain of life" was controversial at the time, and it would be nearly a decade before Woese's interpretation of his molecular microbiology work would be widely accepted by the scientific community.

16. C. R. Woese, "On the Evolution of Cells." *Proceedings of the National Academy of Sciences* 99, no. 13, 2002, 8742–8747, https://doi.org/10.1073/pnas.132266999. HGT is also known as lateral gene transfer.

17. "Soon" being relative in terms of cosmological time. Such an emergence may be a handful of decades away, or it may be many millennia. I'd bet on the nearer term, but either way it's an eye-blink in the history of our world and the universe.

18. Manuel Leal and Brian J. Powell, "Behavioural Flexibility and Problem-Solving in a Tropical Lizard," *Biology Letters* 8, no. 1, April 20, 2011, http://doi.org/10.1098/rsbl.2011.0480.

19. Monica Gagliano, "The Mind of Plants: Thinking the Unthinkable." *Communicative & Integrative Biology*, 10:2. 2017. doi: 10.1080/19420889.2017.1288333.

Anthony Trewavas, "Intelligence, Cognition, and Language of Green Plants." *Frontiers in Psychology* 7, 588, April 26, 2016, doi:10.3389/fpsyg.2016.00588.

20. BCE – Before Common Era, a scientific convention.

21. Rochelle M. Soo, "On the Origins of Oxygenic Photosynthesis and Aerobic Respiration in Cyanobacteria," *Science*, American Association for the Advancement of Science, March 31, 2017, science.sciencemag.org/content/355/6332/1436.

22. Katia Moskvitch, "Slime Molds Remember—But Do They Learn?" *Quanta Magazine*, July 9, 2018, www.quantamagazine.org/slime-molds-remember-but-do-they-learn-20180709/.

A. Boussard, et al., "Memory Inception and Preservation in Slime Moulds: The Quest for a Common Mechanism," *Philosophical Transactions of the Royal Society B*, royalsocietypublishing.org/doi/10.1098/rstb.2018.0368.

23. Thin appendages that whip back and forth, propelling the cell through liquid mediums.

24. Richard Ellis Hudson, et al., "Altruism, Cheating, and Anticheater Adaptations in Cellular Slime Molds." *The American Naturalist* 160, no. 1, 2002, 31, doi:10.2307/3078996.

Chapter 5

1. Melanie Mitchell, *Complexity: A Guided Tour* (New York: Oxford University Press, 2011).
2. As opposed to abstract emergences such as those that occur in computer simulations.
3. Samuel A Ocko, et al., "Solar-Powered Ventilation of African Termite Mounds," *Journal of Experimental Biology* 220, no. 18, 2017, 3,260–3,269 . doi: 10.1242/jeb.160895.
4. George F. Young, et al. "Starling Flock Networks Manage Uncertainty in Consensus at Low Cost." *PLoS Computational Biology* 9, no. 1, 2013, doi:10.1371/journal.pcbi.1002894.

    Anna Azvolinsky, "Birds of a Feather . . . Track Seven Neighbors to Flock Together." *Princeton University*, The Trustees of Princeton University, www.princeton.edu/news/2013/02/07/birds-feather-track-seven -neighbors-flock-together.
5. Craig W. Reynolds, "Flocks, Herds and Schools: A Distributed Behavioral Model," *Proceedings of the 14th Annual Conference on Computer Graphics and Interactive Techniques*, v.21 n.4, 25–34, July 1987, doi:10.1145/37402.37406.
6. Martin Gardner, "Mathematical Games—The Fantastic Combinations of John Conway's New Solitaire Game 'Life'," *Scientific American*, October 1970, 223 (4): 120–123. https://web.stanford.edu/class/sts145/Library /life.pdf.
7. The grid is made infinite by wrapping it back on itself, allowing the cells on what would otherwise be the ends to have the requisite two neighbors.
8. Stephen Wolfram, *A New Kind of Science* (Wolfram Media: 2002), 1,179 . ISBN 978-1-57955-008-0.
9. This ignores the limitations of finite memory in all real-life systems. The original Turing machine is an abstract concept that uses an infinite memory tape.
10. Matthew Cook, "Universality in Elementary Cellular Automata," *Complex Systems* 15, 1–40, 2004. https://www.complex-systems.com /abstracts/v15_i01_a01/.

Wolfram, Stephen (2002). *A New Kind of Science.* Wolfram Media. ISBN 1-57955-008-8.

11. Langton's calculation actually results in a value between zero and one with those lambda values above 0.50 being mirror images, diminishing in complexity as they approach a value or 1.0. This results in zero and one being equally static and 0.273 and 0.727 being equally complex.

12. The Bak-Tang-Wiesenfeld sandpile model was the first discovered system displaying self-organized criticality.

13. At standard temperature and pressure (STP).

14. Per Bak, et al., "Self-Organized Criticality: An Explanation of the 1/f noise," *Physical Review Letters* 59, no. 4, 1987, 381–384. doi:10.1103/PhysRevLett.59.381.

15. Henrik Jjeldtoft Jensen, "Self-Organized Criticality: Emergent Complex Behavior in Physical and Biological Systems," *Physics Today* 52 (10), December 1998, doi: 10.1063/1.882869.

16. Jennifer Ouellette, "Rat Brains Provide Even More Evidence Our Brains Operate near Tipping Point," *Ars Technica*, June 7, 2019, arstechnica.com/science/2019/06/does-the-human-brain-teeter-on-the-edge-of-chaos-rat-brains-point-to-yes/.

    Antonio J. Fontenele, et al., "Criticality between Cortical States," *Physical Review Letters*, 122, 208101, 2018, doi:10.1101/454934.

17. A trend that is itself an emergent manifestation of probability and statistical mechanics.

18. Though criticality can always come into play, wiping out a species, potentially even us, before continuing on the path of growing complexity.

19. Christopher J. Conselice, et al., "The Evolution of Galaxy Number Density at Z < 8 and Its Implications," *Astrophysical Journal* 830, no. 2, 2016, doi:10.3847/0004-637x/830/2/83.

Chapter 6

1. Shannon P. McPherron, et al., "Evidence for Stone-Tool-Assisted Consumption of Animal Tissues before 3.39 Million Years Ago at Dikika, Ethiopia," *Nature* 466, August 12, 2010, 857–860, doi.org/10.1038/nature09248.

    Sonia Harmond, et al., "3.3-Million-Year-Old Stone Tools from Lomekwi 3, West Turkana, Kenya," *Nature* 521, May 21, 2015, 310–315. https://doi.org/10.1038/nature14464.

2. Dietrich Stout and Nada Khreisheh, "Skill Learning and Human Brain Evolution: An Experimental Approach," *Cambridge Archaeological Journal* 25: 867–875 (2015), doi.org/10.1017/S0959774315000359.

    Dietrich Stout, "Tales of a Stone Age Neuroscientist," *Scientific American*, April 2016, https://doi.org/10.1038/scientificamerican0416-28.

3. Flavia Venditti, et al., "Recycling for a Purpose in the Late Lower Paleolithic Levant: Use-Wear and Residue Analyses of Small Sharp Flint Items Indicate a Planned and Integrated Subsistence Behavior at Qesem Cave (Israel)," *Journal of Human Evolution*, 2019; 131: 109 doi: 10.1016/j.jhevol.2019.03.016.

4. Steven Pinker, *The Language Instinct: How the Mind Creates Language* (PLACE: William Morrow, 1994).

5. Wolfgang Enard, et al., "Molecular Evolution of FOXP2, a Gene Involved in Speech and Language." *Nature* 418, August 22, 2002, 869–872. https://doi.org/10.1038/nature01025.

6. PET: Positron emission tomography; MRI: Magnetic resonance imaging; DTI: Diffusion tensor imaging.

7. Exaptation is the co-opting of a previously evolved trait for a different use or trait.

8. Ignacio Martínez, "Human Hyoid Bones from the Middle Pleistocene Site of the Sima De Los Huesos (Sierra De Atapuerca, Spain)," *Journal of Human Evolution*, Academic Press 54, issue 1, January 2008, 118–124, https://doi.org/10.1016/j.jhevol.2007.07.006.

9. John Tooby and Irven DeVore, "The Reconstruction of Hominid Evolution through Strategic Modeling" in *The Evolution of Human Behavior: Primate Models*, ed. Warren G. Kinzey (Albany: SUNY Press, 1987).

10. Steven Pinker, "Planet of the Humans: The Leap to the Top." World Science Festival. 2015. https://youtu.be/ubZ3d-g2lUc

11. The First Agricultural Revolution is also known as the Neolithic Revolution.

12. An accurate count was never fully determined, as McCarthy lost his list of invited participants and visitors. A preliminary replacement list of forty-one people was reassembled, but many didn't turn up or had to cancel. Mathematician and computer scientist Trenchard More from the University of Rochester, whose own notes provided major documentation of the workshop, submitted a list of thirty-two, but Solomon's list of twenty may be the most accurate assembly of definite attendees.

13. Alfred North Whitehead and Bertrand Russell, *Principia Mathematica* (Cambridge University Press, 1910).

14. Boole was also the great-great grandfather of AI giant Geoffrey Hinton, whose own work with neural networks had an enormous impact on twenty-first-century computing.

15. "Learned men have long since thought of some kind of language or universal characteristic by which all concepts and things can be put into beautiful order." G. Leibniz, *On the General Characteristic*, 1679.

16. James Lighthill, "Artificial Intelligence: A General Survey" in *Artificial Intelligence: A Paper Symposium*, Science Research Council, 1973. http://www.chilton-computing.org.uk/inf/literature/reports/lighthill_report/p001.htm.

17. Based on a simplified model of a neuron, perceptrons hinted at the potential of neural networks but would need further advances before they could become truly useful.

18. Marvin Minsky and Seymour A. Papert, *Perceptrons* (Cambridge, MA: MIT Press, 1969).

19. Paul Werbos, "Beyond Regression: New Tools for Prediction and Analysis in the Behavioral Sciences," PhD thesis, Harvard University. 1974.

20. Geoffrey E. Hinton and Ruslan Salakhutdinov, "Reducing the Dimensionality of Data with Neural Networks." *Science* 28 313:504–507. 2006. doi: 10.1126/science.1127647.

    Ruslan Salakhutdinov and Geoffrey E. Hinton, G. E. Learning a non-linear embedding by preserving class neighbourhood structure. In M. Meila & X. Shen (Eds.), Proceedings of the International Conference on Artificial Intelligence and Statistics, vol. 11, 412–419, Cambridge, MA: MIT Press, 2007.

21. Dario Amodei and Danny Hernandez, "AI and Compute." *OpenAI*, March 7, 2019, openai.com/blog/ai-and-compute.

22. The ILSVRC (ImageNet Large-Scale Visual Recognition Challenge) is an annual competition between image classification and object recognition algorithms for large-scale image indexing.

23. John Hawks, et al., "Recent Acceleration of Human Adaptive Evolution." *Proceedings of the National Academy of Sciences* 104, no. 52, 2007, 20,753–20,758. https://doi.org/10.1073/pnas.0707650104.

Chapter 7

1. Yaniv Leviathan and Yossi Matias, "Google Duplex: An AI System for Accomplishing Real-World Tasks Over the Phone," *Google AI Blog*, May 8, 2018. https://ai.googleblog.com/2018/05/duplex-ai-system-for-natural-conversation.html.
2. Alan M. Turing, "Computing Machinery and Intelligence," *Mind* LIX, issue 236, October 1950, 433–460, doi.org/10.1093/mind/LIX.236.433.
3. The Imitation Game has come to be reserved for one specific version of the Turing test, of which there are at least three.
4. Joseph Weizenbaum, *Computer Power and Human Reason: From Judgment to Calculation* (San Francisco: W. H. Freeman, 1976), 7.
5. Byron Reeves and Clifford Nass, *The Media Equation: How People Treat Computers, Television, and New Media Like Real People and Places* (Cambridge University Press, 1996).
6. Brenda Laurel, ed., *The Art of Human-Computer Interface Design* (New York: Addison-Wesley, 1990).
7. "Smart Homes: Vendor Analysis, Impact Assessments & Strategic Opportunities 2018–2023," Juniper Research, 2018.
8. "Impressive Numbers and Stats From Chatbot Dominance So Far," *SmartMessage*, May 27, 2019, www.smartmessage.com/impressive-numbers-stats-chabot-dominance-far/.
9. "Chatbot Conversations to Deliver $8 Billion in Cost Savings by 2022," Juniper Research, July 24, 2017, https://www.juniperresearch.com/analyst xpress/july-2017/chatbot-conversations-to-deliver-8bn-cost-saving.
10. "Chatbot Market Size to Reach $1.25 Billion by 2025," Grand View Research, August 2017, https://www.grandviewresearch.com/press-release/global-chatbot-market.

Chapter 8

1. Irving Biederman. "Recognition by Components: A Theory of Human Image Understanding," *Psychological Review*, 94(2), 1987. https://pdfs.semanticscholar.org/1e38/9040dbdb3057ff510df13808be153c459fd0.pdf.
2. Li Fei-Fei, Rob Fergus, and Pietro Perona, "A Bayesian Approach to Unsupervised One-shot Learning of Object Categories," *Proceedings of the IEEE International Conference on Computer Vision* (2003) 2, 1,134–1,141. doi.org/10.1109/ICCV.2003.1238476.

3. Gregory Koch, "Siamese Neural Networks for One-shot Image Recognition." PhD dissertation, University of Toronto, 2015.

4. James Kirkpatrick, et al., "Overcoming Catastrophic Forgetting in Neural Networks," *Proceedings of the National Academy of Sciences* 114 (13) 3,521–3,526, March 28, 2017, https://doi.org/10.1073/pnas.1611835114.

    James Kirkpatrick, et al., "Enabling Continual Learning in Neural Networks," *Google DeepMind Blog* March 13, 2017, https://deepmind.com/blog/enabling-continual-learning-in-neural-networks.

5. "DARPA Announces $2 Billion Campaign to Develop Next Wave of AI Technologies," Defense Advanced Research Projects Agency, September 7. 2018, www.darpa.mil/news-events/2018-09-07.

6. AI Next Campaign, DARPA, September 7, 2018, https://www.darpa.mil/work-with-us/ai-next-campaign.

7. Doug Lenat, Mayank Prakash, and Mary Shepherd, "CYC: Using Common Sense Knowledge to Overcome Brittleness and Knowledge Acquistion Bottlenecks," *AI Magazine*, vol. 6 no. 4, p. 65–85, 1985, ISSN 0738–4602.

8. Amazon's Mechanical Turk is a crowdsourced method of drawing on human intelligence to perform specific tasks. It is named after an eighteenth-century chess-playing automaton, within which was hidden a human chess master, concealing the true nature of the machine's intelligence.

9. "About Iconary from AI2 | Draw and Guess with AllenAI," Allen Institute for Artificial Intelligence, iconary.allenai.org/about/.

10. Barret Zoph and Quoc V. Le, "Neural Architecture Search with Reinforcement Learning" *ArXiv.org,* November 4, 2016, https://arxiv.org/abs/1611.01578.

11. Chenxi Liu, et al., "Progressive Neural Architecture Search," *ArXiv.org*, 26 July 2018, arxiv.org/abs/1712.00559.

12. Hieu Pham, et al. "Efficient Neural Architecture Search via Parameter Sharing," *ArXiv.org,* February 12, 2018, arxiv.org/abs/1802.03268.

13. Hava Siegelmann, "Lifelong Learning Machines (L2M)," Defense Advanced Research Projects Agency, www.darpa.mil/program/lifelong-learning-machines.

14. Elizabeth S. Spelke and Katherine D. Kinzler, "Core Knowledge," *Developmental Science* 10:1 (2007), 89–96, doi: 10.1111/j.1467-7687.2007.00569.x.

15. Gary Marcus, "Innateness, AlphaZero, and Artificial Intelligence," *ArXiv.org*, January 17, 2018, arxiv.org/abs/1801.05667.

16. Peter Diamandis, "China's BAT: Baidu, Alibaba & Tencent," *Diamandis Tech Blog, China Series*. https://www.diamandis.com/blog/baidu-alibaba -tencent.

17. "Baidu DuerOS Voice Assistant Install Base Doubles in 6 Months to 100 Million." *GlobeNewswire*, August 7, 2018, www.globenewswire.com /news-release/2018/08/07/1548541/0/en/Baidu-DuerOS-Voice -Assistant-Install-Base-Doubles-in-6-Months-to-100-Million.html.

18. Jinxing Yu, et al., "Joint Embeddings of Chinese Words, Characters, and Fine-Grained Subcharacter Components," Proceedings of the 2017 Conference on Empirical Methods in Natural Language Processing, doi: 10.18653/v1/D17-1027.

    Cao, S., Lu, W., Zhou, J., Li, X.: cw2vec: "Learning Chinese Word Embeddings with Stroke N-gram Information." The AAAI Publications, Thirty-Second AAAI Conference on Artificial Intelligence, 158–160, 2018, https://aaai.org/ocs/index.php/AAAI/AAAI18/paper/view/17444.

19. "Putin: Leader in Artificial Intelligence Will Rule World," Associated Press, September 1, 2017, https://apnews.com/bb5628f2a7424a10b3e 38b07f4eb90d4.

Chapter 9

1. Rosalind W. Picard, *Affective Computing* (MIT Press: 1997).

2. Now Kantar Millward Brown.

3. Matt Day, "Amazon Is Working on a Device That Can Read Human Emotions," *Bloomberg*, May 23, 2019, www.bloomberg.com/news /articles/2019-05-23/amazon-is-working-on-a-wearable-device-that -reads-human-emotions.

4. Rana el Kaliouby, "This App Knows How You Feel–from the Look on Your Face," TED.com, https://www.ted.com/talks/rana_el_kaliouby _this_app_knows_how_you_feel_from_the_look_on_your_face.

    el Kaliouby made this statement in 2015, so obviously we will need longer than five years, but the point is made.

5. "Gartner Survey Finds Consumers Would Use AI to Save Time and Money." *Gartner*. September 12, 2018, https://www.gartner.com/en/news room/press-releases/2018-09-12-gartner-survey-finds-consumers -would-use-ai-to-save-time-and-money.

6. "13 Surprising Uses For Emotion AI Technology." *Smarter with Gartner*, https://www.gartner.com/smarterwithgartner/13-surprising-uses-for-emotion-ai-technology/.

7. Picard has estimated that if they had linearly scaled the work they were doing at the Lab at the time, generating the needed training set would have cost them a billion dollars.

8. E. Maor, et al. "Voice Signal Characteristics Are Independently Associated with Coronary Artery Disease." Mayo Clin Proc. 2018, 93: 840–847a, doi: 10.1016/j.mayocp.2017.12.025

9. "IDC Forecasts Worldwide Technology Spending on the Internet of Things to Reach $1.2 Trillion in 2022." IDC. June 18, 2018, https://www.idc.com/getdoc.jsp?containerId=prUS43994118.

10. Adam Sadilek and John Krumm, "Far Out: Predicting Long-Term Human Mobility," *Proceedings of the 26th AAAI Conference on Artificial Intelligence*, July 2012, https://www.microsoft.com/en-us/research/wp-content/uploads/2016/12/Sadilek-Krumm_Far-Out_AAAI-2012.pdf.

11. Richard Yonck, *Heart of the Machine: Our Future in a World of Artificial Emotional Intelligence* (Arcade Publishing: 2017).

12. Michael W. Eysenck, "Arousal, Learning, and Memory," *Psychological Bulletin* 83(3), May 1976, 389–404. doi:10.1037/0033–2909.83.3.389.

13. Antonio Damasio, *Descartes' Error: Emotion, Reason, and the Human Brain* (Putnam: 1994).

14. Thank you to Oren Etzioni for this evocative illustration.

Chapter 10

1. In the fourth century BC, Aristotle maintained that intelligence originated in the heart.

2. A group of organisms considered to have descended from a common ancestor.

3. There has been some controversy regarding the interpretation of certain areas being fossilized brain and nerve tissue, since these soft features are likely to deteriorate too rapidly to be fossilized. An alternate interpretation of radiating bacterial biofilms has been proposed.

4. Blue Brain Project, https://www.epfl.ch/research/domains/bluebrain/.

5. Human Brain Project, https://www.humanbrainproject.eu/en/.

6. Exascale supercomputers will have sustained processing speed on the order of exaflops at 10^{18} (a quintillion) floating point operations per

second. This is 10,000 times faster than the Blue Gene supercomputer in 2005 and a thousand times more powerful than the first petaflop supercomputer built at Los Alamos in 2008. Several nations are working on exascale projects and many are anticipated to come online in 2020 and 2021.

7. To date, there is no formal definition of cognitive computing, which can be a system that mimics the function of the brain or a system that assists in decision-making.

8. James Randerson, "How Many Neurons Make a Human Brain? Billions Fewer than We Thought," *Guardian*, February 28, 2012, amp.theguardian. com/science/blog/2012/feb/28/how-many-neurons-human-brain.

9. "Largest Neuronal Network Simulation Achieved Using K Computer." *RIKEN*, August 2, 2013, www.riken.jp/en/pr/press/2013/20130802_1.

10. "Neuromorphic Computing Breaks New Ground in Brain Simulation." *TOP500 Supercomputer Sites*, www.top500.org/news/neurmorphic -computing-breaks-new-ground-in-brain-simulation.

11. Johannes Schemmel, et al., "An Accelerated Analog Neuromorphic Hardware System Emulating NMDA and Calcium-based Non-linear Dendrites," 2017, *International Joint Conference on Neural Networks (IJCNN)*, 2,217–2226. doi: 10.1109/IJCNN.2017.7966124

12. After the legendary mathematician and physicist John von Neumann, who described the concept in a 1945 report.

13. Michael S. Gazzaniga, *Consciousness Instinct: Unraveling the Mystery of How the Brain Makes the Mind* (New York: Farrar, Straus & Giroux, 2019).

14. Marvin Minsky, *The Society of Mind* (New York: Simon and Schuster, 1986).

15. Marvin Minsky, *The Emotion Machine: Commonsense Thinking, Artificial Intelligence, and the Future of the Human Mind* (New York: Simon & Schuster, 2007).

16. Olaf Sporns, "Olaf Sporns on Network Neuroscience," *Network Neuroscience*, MIT Press podcast, April 25, 2018, https://mitpress.podbean .com/e/olaf-sporns-on-network-neuroscience.

17. Michael Gazzaniga, *Who's In Charge: Free Will and the Science of the Brain*, New York: HarperCollins, 2011.

Chapter 11

1. "Brain Chip Reads Man's Thoughts," *BBC News*, March 31, 2005, http://news.bbc.co.uk/2/hi/health/4396387.stm.

2. Leigh R. Hochberg, et al., "Neuronal Ensemble Control of Prosthetic Devices by a Human with Tetraplegia," *Nature*, July 13, 2006, 442 (7099): 164–71, https://www.nature.com/articles/nature04970.

3. Renee Meiller, "Researchers Use Brain Interface to Post to Twitter." *University of Wisconsin-Madison News*, April 20, 2009, news.wisc.edu/researchers-use-brain-interface-to-post-to-twitter.

4. Carles Grau, et al., "Conscious Brain-to-Brain Communication in Humans Using Non-Invasive Technologies," *PLoS ONE*, 2014; 9 (8): e105225, doi: 10.1371/journal.pone.0105225.

5. Doree Armstrong and Michelle Ma, "Researcher Controls Colleague's Motions in First Human Brain-to-Brain Interface," *University of Washington News*, www.washington.edu/news/2013/08/27/researcher-controls-colleagues-motions-in-1st-human-brain-to-brain-interface.

6. Larry Hardesty, "Computer System Transcribes Words Users 'Speak Silently,'" *MIT News*, April 4, 2018, news.mit.edu/2018/computer-system-transcribes-words-users-speak-silently-0404.

7. Charles Jorgensen, et al., "Sub Auditory Speech Recognition Based on EMG Signals," *Proceedings of the International Joint Conference on Neural Networks, IEEE*, 2003, https://ieeexplore.ieee.org/document/1224072.

8. Hassan Akbari, et al., "Towards Reconstructing Intelligible Speech from the Human Auditory Cortex," *BioRxiv*, Cold Spring Harbor Laboratory, January 1, 2018, www.biorxiv.org/content/10.1101/350124v2.

9. Hassan Akbari, et al., "Towards Reconstructing Intelligible Speech from the Human Auditory Cortex." *Nature Scientific Reports* 9, article number 874, January 29, 2019, https://www.nature.com/articles/s41598-018-37359-z.

10. An important consideration is that this study is not translating abstract thoughts into speech, but rather the speech activation signals. This is similar to activating subvocalization technology, only with the signals originating deeper in the brain.

11. The BRAIN Initiative, NIH: https://braininitiative.nih.gov.

12. Al Emondi, "Systems-Based Neurotechnology for Emerging Therapies (SUBNETS)," DARPA, https://www.darpa.mil/program/systems-based-neurotechnology-for-emerging-therapies.

13. Patrick Tucker, "The Military Is Building Brain Chips to Treat PTSD." *Defense One*, May 28, 2014, https://www.defenseone.com/technology/2014/05/D1-Tucker-military-building-brain-chips-treat-ptsd/85360.

14. Tristan McClure-Begley, "Restoring Active Memory (RAM)," DARPA, https://www.darpa.mil/program/restoring-active-memory.

15. David Eagleman, Facebook post, March 31, 2017, https://www.facebook.com/David.M.Eagleman/posts/a-great-deal-of-interest-has-recently-blossomed-regarding-futuristic-ways-to-rea/10155125223381549.

16. Bryan Johnson, "What If: The Next Frontier of Human Aspiration" (transcript), Code Conference 2017, Rancho Palos Verdes, CA, May 31, 2017, https://bryanjohnson.co/next-frontier-human-aspiration.

17. Bryan Johnson, "Kernel's Quest to Enhance Human Intelligence," Medium, October 20, 2016, https://medium.com/@bryan_johnson/kernels-quest-to-enhance-human-intelligence-7da5e16fa16c.

Chapter 12

1. Dan Goodin, "Insulin Pump Hack Delivers Fatal Dosage over the Air," *The Register*, October 31, 2011, https://www.theregister.co.uk/2011/10/27/fatal_insulin_pump_attack/.

2. Julia Shaw and Stephen Porter. "Constructing Rich False Memories of Committing Crime," *Psychological Science* 26, No. 3, 2015, 291–301, https://doi.org/10.1177/0956797614562862.

3. Tamara Bonaci, et al., "App Stores for the Brain: Privacy and Security in Brain–Computer Interfaces," *IEEE Technology and Society Magazine*, 2015; 34(2): 32–39. doi: 10.1109/MTS.2015.2425551.

4. Marco Iacoboni, *Mirroring People: The New Science of How We Connect with Others* (New York: Picador, 2008).

5. A nuclease is an enzyme that cleaves the nucleotides in RNA and DNA, cutting them into smaller units.

6. F. Chollet, "What Worries Me about AI," Medium, March 28, 2018, https://medium.com/@francois.chollet/what-worries-me-about-ai-ed9df07 2b704.

Chapter 13

1. This, despite the fact there are many different singularities in the world of mathematics, physics, cosmology, and other fields, as well as multiple interpretations of the technological singularity itself.

2. Its processing power relative to its size, energy requirements, and cost.

3. About 1,200 feet across, it's more of a small lake.

4. Amara was president of the Institute for the Future, a Palo Alto–based think tank spun off from the Rand Corporation in 1968.

---

5. Bill Gates, et al., *The Road Ahead*, New York: Penguin Books, 1995.

6. Though the term "explosion" may seem harsh, this process strongly resembles the runaway reaction seen in nuclear bombs.

7. Waldemar Kaempffert, "Rutherford Cools Atom Energy Hope," *New York Times*, September 12, 1933.

8. Szilard reputedly conceived the chain reaction the next day while crossing a street in London.

9. Vernor Vinge, "The Coming Technological Singularity," *Whole Earth Review*, Winter 1993.

10. Vernor Vinge, "Vernor Vinge SIAI Interview," *SIAI*, 2008, https://youtu.be/IpUKh4thvK0.

11. Nick Bostrom, *Superintelligence: Paths, Dangers, Strategies* (Oxford University Press, 2014), 259.

12. Rory Cellan-Jones, "Stephen Hawking Warns Artificial Intelligence Could End Mankind," *BBC News*, December 2, 2014, https://www.bbc.com/news/technology-30290540.

13. Matt McFarland, "Elon Musk: 'With Artificial Intelligence We Are Summoning the Demon'," *The Washington Post*, October 24, 2014, https://www.washingtonpost.com/news/innovations/wp/2014/10/24/elon-musk-with-artificial-intelligence-we-are-summoning-the-demon/.

14. Of which twenty-nine of the one hundred responded.

15. James Barrat and Ben Goertzel, "How Long Till AGI?—Views of AGI-11 Conference Participants," *H+ Magazine*, September, 16, 2011, https://hplusmagazine.com/2011/09/16/how-long-till-agi-views-of-agi-11-conference-participants/.

16. Martin Ford, *Architects of Intelligence: The Truth about AI from the People Building It* (Birmingham, UK: Packt Publishing, 2018).

17. Stephen M. Omohundro, "The Basic AI Drives," *Proceedings of the First AGI Conference* 171, 483–492. *Frontiers in Artificial Intelligence and Applications*, ed. Wang, P., Goertzel, B., Franklin, S. IOS Press, February 2008, https://www.researchgate.net/publication/221328949_The_basic_AI_drives.

18. In economics, rational behavior refers to a decision-making process that leads to optimal levels of benefit or utility to an agent.

19. Kevin Kelly, *What Technology Wants* (London: Penguin, 2011).

20. The recent short-lived global moratorium on genetically modifying embryos attests to this.

21. Stuart Russell, "A Brave New World?" World Economic Forum, Davos, January 22, 2015, https://www.weforum.org/events/world-economic-forum-annual-meeting-2015/sessions/brave-new-world.

22. A normative ethical theory is the philosophical study of how a person should act ethically.

23. Universal and personal.

24. Ben Goertzel, "Should Humanity Build a Global AI Nanny to Delay the Singularity Until It's Better Understood?" *Journal of Consciousness Studies* 19, 1–2, 2012, 96–111, http://citeseerx.ist.psu.edu/viewdoc/download?doi=10.1.1.352.3966&rep=rep1&type=pdf.

25. Eliezer Yudkowsky, "Creating Friendly AI 1.0: The Analysis and Design of Benevolent Goal Architectures," *MIRI*, 2001, intelligence.org/files/CFAI.pdf.

Chapter 14

1. The Borg are a hive mind society of emotionless human-machine hybrids that have been stripped of independent will. They were introduced in the television series *Star Trek: The Next Generation* in 1989.

2. Futurist Ray Kurzweil has calculated that given the pace of exponential change, we will see 20,000 years of progress during just the twenty-first century. This is based on a standard year progressing at the rate of change observed in the year 2000.

3. Intelligence quotients are controversial measures of intelligence, not least because a single number is used to try to describe a highly multi-faceted attribute.

4. Megan Molteni, "Now You Can Sequence Your Whole Genome for Just $200." *Wired*, November 19, 2018, https://www.wired.com/story/whole-genome-sequencing-cost-200-dollars.

5. Suzanne Sniekers, et al., "Genome-wide Association Meta-analysis of 78,308 Individuals Identifies New Loci and Genes Influencing Human Intelligence," *Nature Genetics*, 2017; doi: 10.1038/ng.3869.

6. Joyce C. Harper, "Introduction to preimplantation genetic diagnosis," in Joyce C. Harper, ed., *Preimplantation Genetic Diagnosis: Second Edition* (New York: Cambridge University Press, 2009).

7. Julian Borger and James Meek, "Parents Create Baby to Save Sister," *Guardian*, October 4, 2000, www.theguardian.com/science/2000/oct/04/genetics.internationalnews.

8. "711th Human Performance Wing," Wright-Patterson AFB, www.
    wpafb.af.mil/afrl/711hpw.
9. Emma Young, "Brain Stimulation: The Military's Mind-zapping Project,"
    BBC Future, June 3, 2014, http://www.bbc.com/future/story/20140603
    -brain-zapping-the-future-of-war.
10. John Woolfolk, "Fremont Family Upset That Kaiser Let 'Robot'
    Deliver Bad News," *Mercury News*, March 9, 2019. www.mercurynews
    .com/2019/03/08/fremont-family-upset-that-kaiser-let-robot-deliver
    -bad-news.

Chapter 15
1. Digital twin has come to refer to all manner of virtual duplicates of real-
    world objects, processes, environments, and people, independent of any
    intelligence. For the purposes of this book, cognitive twin is meant to
    mean a virtual entity with all of a real person's knowledge, though it is
    not necessarily represented as that person. A digital double is essentially
    a cognitive twin that is also a physical representation of the original
    individual.
2. Much like the difference between an identical twin traveling at relativ-
    istic speeds, somehow observing their sibling back on Earth. According
    to Einstein's special theory of relativity, time would slow down dramat-
    ically for the traveling twin.
3. David Eagleman, "Silicon Immortality: Downloading Consciousness
    into Computers," *This Will Change Everything: Ideas That Will Shape the
    Future*, John Brockman, ed. (New York: Harper Perennial, 2010).
4. Air-channel transistors and 2-D semiconductors being only two such
    possibilities.
5. Actually, 230 equals 1,073,741,824.
6. Heraclitus lived several centuries before Plutarch. The paradoxical ship
    was a frequent topic of discussion throughout the centuries by Plato and
    others as well.
7. Because the dentate gyrus is instrumental in establishing episodic mem-
    ory, it's been speculated that the imperfect replacement of these cells
    contributes to the shifting and fading of our memories over the years.
8. Paul Nelson and Joanna Masel, "Intercellular Competition and the
    Inevitability of Multicellular Aging," *Proceedings of the National Academy
    of Sciences*, National Academy of Sciences, December 5, 2017, 114 (49):
    12982–12987, www.pnas.org/content/114/49/12982.

Chapter 16

1. Thomas G. Dietterich, Twitter Post, January 7, 2018, 11:45 p.m., https://twitter.com/tdietterich/status/950272241106804737.

2. Kevin Kelly, *The Inevitable: Understanding the 12 Technological Forces That Will Shape Our Future* (New York: Penguin Books, 2017).

3. Sandbox: a term from computing in which code is isolated from an active environment in order to test it thoroughly before deployment.

4. Orchestrated objective reduction (Orch-OR) is a theory of consciousness proposed by Stuart Hameroff and Roger Penrose that posits consciousness arising from the quantum fluctuations within microtubules in neurons. As Tegmark points out, wave function collapse would occur far too rapidly to influence neural processes.

5. Christiane Schreiweis, et al. "Humanized Foxp2 Accelerates Learning by Enhancing Transitions from Declarative to Procedural Performance," *Proceedings of the National Academy of Sciences*, September 30, 2014, 111(39):14,253–8, https://www.pnas.org/content/111/39/14253.

6. Robert E. Hampson, et al., "Facilitation and restoration of cognitive function in primate prefrontal cortex by a neuroprosthesis that utilizes minicolumn-specific neural firing," *Journal of Neural Engineering*, 2012, 9 (5): 056012, doi: 10.1088/1741–2560/9/5/056012.

7. Xiaoning Han, et al., "Forebrain Engraftment by Human Glial Progenitor Cells Enhances Synaptic Plasticity and Learning in Adult Mice," *Cell Stem Cell* 12, issue 3, 342–353, March 7, 2013, doi: 10.1016/j.stem.2012.12.015.

8. Consider an imaginary conversation between two single-celled organisms in which they debate the merits of one day evolving into multicellular organisms (or nerve cells into minds). From each perspective, it's a no-brainer.

Chapter 17

1. A star having comparable characteristics to our own sun.

2. Special relativity states that as a body travels faster, its mass increases. At nonrelativistic speeds this is virtually nonexistent. Even at 10 percent of light speed, a moving body's mass only increases by about half a percent. But if you could accelerate a mass to 90 percent of light speed, it would double. The rate of increase climbs even more rapidly as the theoretical mass approaches light speed. Since even the smallest particle having mass would become infinite if it could ever reach light speed,

this is considered impossible according to our current understanding of physics.

3. The Anthropocene, the epoch of modern man, has reputedly been responsible for accelerating the extinction rate by a factor of between 100 and 1,000.

4. Curtis W. Marean, "When the sea saved humanity," *Scientific American*, November 1, 2010.

5. Robert A. DePalma, et al., "A Seismically Induced Onshore Surge Deposit at the KPg Boundary, North Dakota," *Proceedings of the National Academy of Sciences*, 2019, 201817407. https://www.pnas.org/content/116/17/8190.

6. Informally known as Eagleworks Laboratories.

7. The quantum vacuum state is also known as zero point field. The zero point energy that results from this field is not completely fixed and can fluctuate as virtual particles pop in and out of existence, due to Heisenberg's uncertainty principle.

8. Robert A. Freitas Jr., "Exploratory Design in Medical Nano-technology: A Mechanical Artificial Red Cell," *Artificial Cells, Blood Substitutes, and Biotechnology*, 26 (1998), 411–430, https://doi.org/10.3109/10731199809117682.

9. Some calculations put this closer to a million years.

10. Martin G. T. A. Rutten, et al. "Encoding Information into Polymers," *Nature Reviews Chemistry* 2, no. 11, 2018, 365–381, doi:10.1038/s41570-018-0051-5.

11. The philosophical idea that any random sample is typical of the most probable category.

12. Recent research indicates long-term exposure to microgravity causes astronauts' brains to expand and fill with fluid, as well as to age more rapidly. Angelique Van Ombergen, et al., "Brain Ventricular Volume Changes Induced by Long-Duration Spaceflight," *Proceedings of the National Academy of Sciences*, 2019, 201820354, https://doi.org/10.1073/pnas.1820354116.

Chapter 18

1. Jan Hilgevoord and Jos Uffink, "The Uncertainty Principle," *Stanford Encyclopedia of Philosophy*, Stanford University, July 12, 2016, https://plato.stanford.edu/entries/qt-uncertainty/.

Panu Raatikainen, "Gödel's Incompleteness Theorems," *Stanford Encyclopedia of Philosophy*, Stanford University, January 20, 2015, https://plato.stanford.edu/entries/goedel-incompleteness/.

2. "2015 International Technology Roadmap for Semiconductors (ITRS)," Semiconductor Industry Association, www.semiconductors.org/resources/2015-international-technology-roadmap-for-semi-conductors-itrs.

3. The Landauer limit states the absolute minimum energy needed to perform a computer processing cycle.

4. Many additions have been suggested to the original scale, most prominently the inclusion of 0, IV, and V civilizations.

5. Type 0 was calculated by Carl Sagan to be a pretechnological planet that controlled 1 megawatt of power.

6. Michio Kaku, "The Physics of Interstellar Travel: To One Day Reach the Stars," Blog post, MKaku.org, 2010. https://mkaku.org/home/articles/the-physics-of-interstellar-travel.

7. Who in turn was inspired by Olaf Stapledon's 1937 science fiction novel *The Star Maker*.

8. Anders Sandberg, et al., "Dyson Sphere FAQ," https://www.aleph.se/Nada/dysonFAQ.html.

9. A planetesimal is a gravitationally bound body that is a small fraction the size of a planet—usually a minimum of one kilometer in diameter.

10. Yigit Dallilar, et al., "A Precise Measurement of the Magnetic Field in the Corona of the Black Hole Binary V404 Cygni," *Science* 358, issue 6368, 1,299–1,302 , December 8, 2017, doi: 10.1126/science.aan0249.

11. Roger L. Griffith, et al., "The Ĝ Infrared Search For Extraterrestrial Civilizations With Large Energy Supplies. The Reddest Extended Sources In WISE," The Astrophysical Journal Supplement Series 217, no. 2, 2015, https://iopscience.iop.org/article/10.1088/0067–0049/217/2/25.

12. Robert J. Bradbury, "Matrioshka Brains," 1997–2000, https://www.gwern.net/docs/ai/1999-bradbury-matrioshkabrains.pdf.

13. M. A. Garrett, "Application of the Mid-IR Radio Correlation to the Ĝ Sample and the Search for Advanced Extraterrestrial Civilisations," *Astronomy & Astrophysics,* 581, 2015, doi:10.1051/0004–6361/201526687.

14. Anders Sandberg, "The Physics of Information Processing Superobjects: Daily Life Among the Jupiter Brains," *Journal of Evolution and Technology*, 5(1), 1–34, December 22, 1999.

15. Q. Y. Liang, et al., "Observation of Three-Photon Bound States in a Quantum Nonlinear Medium," *Science* 359, no. 6377, 2018, 783–786, doi:10.1126/science.aao7293.

Chapter 19

1. Unknowable because the universe is so much larger than 100 trillion light-years across at this point, that the information from most parts of it will never have time to reach us.
2. Neil deGrasse Tyson sometimes jokingly refers to dark matter as Fred, because we really don't know what it is.
3. Yes, a few of the secrets of time travel have been worked out by now; I'm just not saying which ones.
4. A tip of the hat to Douglas Adams, Spider Robinson, and George Lucas.
5. Including theoretical physicist James Gates, astrophysicist Neil deGrasse Tyson, philosopher Nick Bostrom, physicist Max Tegmark, and entrepreneur Elon Musk.
6. "How Facebook Researchers' Realistic Simulations Help Advance AI and AR." Facebook Technology, June 14, 2019, tech.fb.com/facebook -reality-labs-replica-simulations-help-advance-ai-and-ar.
7. Silas R. Beane, et al., "Constraints on the Universe as a Numerical Simulation," *The European Physical Journal* A 50, no. 9, 2014, doi:10.1140 /epja/i2014-14148-0.
8. Zohar Ringel and Dmitry L. Kovrizhin. "Quantized Gravitational Responses, the Sign Problem, and Quantum Complexity," *Science Advances* 3, no. 9, 2017, doi:10.1126/sciadv.1701758.
9. Nobuyuki Sakai, et al., "Is It Possible to Create a Universe out of a Monopole in the Laboratory?" *Physical Review D* 74, no. 2, 2006, doi:10.1103/physrevd.74.024026.
10. A Grand Unified Theory (GUT) is a high-energy particle physics model that seeks to unify the strong, weak, and electromagnetic forces into a single force. This would have been the state of our universe at the very earliest stages of the Big Bang.
11. Some versions of these hypotheses theorize that the physical laws in a monopole universe would be the same as in our own.
12. By many orders of magnitude.

INDEX